ARM® Cortex® M4 Cookbook

Over 50 hands-on recipes that will help you develop amazing real-time applications using GPIO, RS232, ADC, DAC, timers, audio codecs, graphics LCD, and a touch screen

Dr. Mark Fisher

[PACKT]
PUBLISHING

BIRMINGHAM - MUMBAI

ARM® Cortex® M4 Cookbook

First published: March 2016

Production reference: 1020316

Published by Packt Publishing Ltd.
Livery Place
35 Livery Street
Birmingham B3 2PB, UK.

ISBN 978-1-78217-650-3

www.packtpub.com

Credits

Author
Dr. Mark Fisher

Reviewer
Alex Barrett

Commissioning Editor
Kunal Parikh

Acquisition Editor
Sonali Vernekar

Content Development Editor
Mayur Pawanikar

Technical Editor
Kunal Chaudhari

Copy Editors
Priyanka Ravi
Sonia Mathur

Project Coordinator
Nidhi Joshi

Proofreader
Safis Editing

Indexer
Monica Ajmera Mehta

Graphics
Disha Haria

Production Coordinator
Nilesh Mohite

Cover Work
Nilesh Mohite

About the Author

Dr. Mark Fisher is a chartered engineer, MIET. He started his career as an electronics apprentice with the UK Ministry of Defence. This was before he studied Electrical and Electronic Engineering at Aston University, Birmingham. After his graduation, he joined Ferranti Computer Systems, Manchester. However, he returned to academia to study Microprocessor Engineering and Digital Electronics at Manchester University (UMIST), and he then remained as a research assistant within the Department of Computation to gain a PhD in Applied Machine Learning. Currently, he is a senior lecturer at the School of Computing Sciences, University of East Anglia, and the course director of the Computer Systems Engineering Degree programme. Many of the recipes in this book were originally developed in the context of a taught module that Mark leads, which is popular among undergraduate and master's students in the school.

Mark currently researches in the fields of medical imaging and computer vision, and he is a co-author of over a hundred journal and conference papers in this area.

About the Reviewer

Alex Barrett has been heavily involved in all aspects of design, development, and manufacture of electronic systems and devices as a director of design consultants Rocolec Ltd. for over twenty years. Prior to this, he worked in the oil industry in the designing and manufacturing of remotely-operated submersible vehicles (ROVs), and manufacturing and testing television reception equipment. He enjoys traveling, and he has an interest in languages, currently focusing on learning Russian. He is also a volunteer on the Anglian Coastal committee of The Institution of Engineering and Technology.

www.PacktPub.com

eBooks, discount offers, and more

Did you know that Packt offers eBook versions of every book published, with PDF and ePub files available? You can upgrade to the eBook version at www.PacktPub.com and as a print book customer, you are entitled to a discount on the eBook copy. Get in touch with us at customercare@packtpub.com for more details.

At www.PacktPub.com, you can also read a collection of free technical articles, sign up for a range of free newsletters and receive exclusive discounts and offers on Packt books and eBooks.

PACKTLiB™

https://www2.packtpub.com/books/subscription/packtlib

Do you need instant solutions to your IT questions? PacktLib is Packt's online digital book library. Here, you can search, access, and read Packt's entire library of books.

Why Subscribe?

- ▶ Fully searchable across every book published by Packt
- ▶ Copy and paste, print, and bookmark content
- ▶ On demand and accessible via a web browser

Table of Contents

Preface

This book begins with an introduction to the ARM Cortex family and covers its basic concepts. We cover the installation of the ARM uVision Integrated Development Environment and topics, such as target devices, evaluation boards, code configuration, and GPIO. You will learn about the core programming topics that deal with structures, functions, pointers, and debugging in this book. You will also learn about various advanced aspects, such as data conversion, multimedia support, real-time signal processing, and real-time embedded systems. You will also get accustomed with creating game applications, programming I/O, and configuring GPIO and UART ports. By the end of this book, you will be able to successfully create robust and scalable ARM Cortex-based applications.

What this book covers

Chapter 1, *A Practical Introduction to ARM® Cortex®*, shows you how to compile, download, and run simple programs on an evaluation board.

Chapter 2, *C Language Programming*, introduces you to writing programs in C, a high-level language that was developed in the 1970s and is popular among embedded-system developers.

Chapter 3, *Programming I/O*, investigates some of the functions that configure I/O devices, and you will gain an understanding of what is involved in writing I/O interfaces for other targets.

Chapter 4, *Assembly Language Programming*, explains how to write functions in assembly language. Assembly language is a low-level programming language that is specific to particular computer architecture. Therefore, unlike programs written high-level languages, programs written in assembly language cannot be easily ported to other hardware architectures.

Chapter 5, *Data Conversion*, introduces approaches to data conversion, namely analog to digital conversion and vice versa. This chapter also covers the principal features used by microcontrollers for data conversion.

Chapter 6, Multimedia Support, discusses support for various multimedia peripherals, which are discrete components connected to the microcontroller by a bus. Support for an LCD touchscreen, audio codec, and camera peripherals is a very attractive feature of the STM32F4xxx microcontroller, and selecting an evaluation board that includes these peripherals, although more expensive, will be covered in this chapter.

Chapter 7, Real-Time Signal Processing, introduces you to Digital Signal Processing (DSP) and reviews the ARM Cortex M4 instruction set support for DSP applications. This chapter will walk through a DMA application using the codec, followed by designing a low-pass filter.

Chapter 8, Real-Time Embedded Systems, shows you how to write a multithreaded program using flags for communication and ensuring mutual exclusion when accessing shared resources.

Chapter 9, Embedded Toolchain, teaches you how to install the GNU ARM Eclipse toolchain for the Windows Operating System and to build and run a simple Blinky program on the MCBSTM32F400 evaluation board. This chapter will also show you how to use the STM32CubeMX Framework (API) and how to port projects to GNU ARM Eclipse.

What you need for this book

You require the Keil Development Board MCBSTM32F400 (v1.1) and ARM ULINK-ME for this book.

Who this book is for

This book is aimed at those with an interest in designing and programming embedded systems. These could include electrical engineers or computer programmers who want to get started with microcontroller applications using the ARM Cortex M4 architecture in a short time frame. This book's recipes can also be used to support students learning embedded programming for the first time. Basic knowledge of programming using a high-level language is essential but those familiar with other high-level languages such as Python or Java should not have too much difficulty picking up the basics of embedded C programming.

Sections

In this book, you will find several headings that appear frequently (Getting ready, How to do it..., How it works..., There's more..., and See also).

To give clear instructions on how to complete a recipe, we use these sections as follows:

Getting ready

This section tells you what to expect in the recipe, and describes how to set up any software or any preliminary settings required for the recipe.

How to do it...

This section contains the steps required to follow the recipe.

How it works...

This section usually consists of a detailed explanation of what happened in the previous section.

There's more...

This section consists of additional information about the recipe in order to make the reader more knowledgeable about the recipe.

See also

This section provides helpful links to other useful information for the recipe.

Conventions

In this book, you will find a number of text styles that distinguish between different kinds of information. Here are some examples of these styles and an explanation of their meaning.

Code words in text, database table names, folder names, filenames, file extensions, pathnames, dummy URLs, user input, and Twitter handles are shown as follows: "Copy the function named `SystemClock_Config()` from the example."

A block of code is set as follows:

```
#ifdef __RTX
extern uint32_t os_time;

uint32_t HAL_GetTick(void) {
  return os_time;
}
#endif
```

New terms and **important words** are shown in bold. Words that you see on the screen, for example, in menus or dialog boxes, appear in the text like this: "Run the program by pressing **RESET** on the evaluation board."

> Warnings or important notes appear in a box like this.

> Tips and tricks appear like this.

Reader feedback

Feedback from our readers is always welcome. Let us know what you think about this book—what you liked or disliked. Reader feedback is important for us as it helps us develop titles that you will really get the most out of.

To send us general feedback, simply e-mail feedback@packtpub.com, and mention the book's title in the subject of your message.

If there is a topic that you have expertise in and you are interested in either writing or contributing to a book, see our author guide at www.packtpub.com/authors.

Customer support

Now that you are the proud owner of a Packt book, we have a number of things to help you to get the most from your purchase.

Downloading the example code

You can download the example code files from your account at http://www.packtpub.com for all the Packt Publishing books you have purchased. If you purchased this book elsewhere, you can visit http://www.packtpub.com/support and register to have the files e-mailed directly to you.

Downloading the color images of this book

We also provide you with a PDF file that has color images of the screenshots/diagrams used in this book. The color images will help you better understand the changes in the output. You can download this file from https://www.packtpub.com/sites/default/files/downloads/ARMCortexM4Cookbook_ColorImages.pdf.

Errata

Although we have taken every care to ensure the accuracy of our content, mistakes do happen. If you find a mistake in one of our books—maybe a mistake in the text or the code—we would be grateful if you could report this to us. By doing so, you can save other readers from frustration and help us improve subsequent versions of this book. If you find any errata, please report them by visiting `http://www.packtpub.com/submit-errata`, selecting your book, clicking on the **Errata Submission Form** link, and entering the details of your errata. Once your errata are verified, your submission will be accepted and the errata will be uploaded to our website or added to any list of existing errata under the Errata section of that title.

To view the previously submitted errata, go to `https://www.packtpub.com/books/content/support` and enter the name of the book in the search field. The required information will appear under the **Errata** section.

Piracy

Piracy of copyrighted material on the Internet is an ongoing problem across all media. At Packt, we take the protection of our copyright and licenses very seriously. If you come across any illegal copies of our works in any form on the Internet, please provide us with the location address or website name immediately so that we can pursue a remedy.

Please contact us at `copyright@packtpub.com` with a link to the suspected pirated material.

We appreciate your help in protecting our authors and our ability to bring you valuable content.

Questions

If you have a problem with any aspect of this book, you can contact us at `questions@packtpub.com`, and we will do our best to address the problem.

1

A Practical Introduction to ARM® CORTEX®

In this chapter, we will cover the following topics:

- ▶ Installing uVision5
- ▶ Linking an evaluation board
- ▶ Running an example program
- ▶ Writing a simple program
- ▶ Understanding the simple use of GPIO
- ▶ Estimating microcontroller performance

Introduction

This chapter will show you how to compile, download, and run simple programs on an evaluation board. A software tool called a **Microcontroller Development Kit** (**MDK**), including an **Integrated Development Environment (IDE)**, is the simplest way of achieving this. Keil (a company owned by ARM) markets an extensive range of software tools to support embedded system development. Amongst these, the MDK-ARM development kit represents an integrated software development environment, supporting devices based on the Cortex-M (and associated) cores (see http://www.keil.com/arm/mdk.asp).

Installing uVision5

A free evaluation version of the IDE known as the MDK-ARM Lite edition, running (albeit with limited functionality) under the Windows operating system, is available for download. The main limitation of the environment is that programs that generate more than 32 KB of code cannot be compiled and linked (see `http://www.keil.com/demo/limits.asp`). However, since most programs written by novices tend be quite small, this limitation is not a serious problem. For those who expect their executable image to exceed 32 KB, other open source compiler and IDE options are considered in *Chapter 9, Embedded Toolchain*.

uVision5, the latest version of the IDE is distributed as two components. An MDK core contains all the development tools, and software packs, together with **Cortex Microcontroller Software Interface standard** (**CMSIS**) and middleware libraries, which add support for target devices.

Installation involves downloading and running an executable (`.exe`) file. Users can download and install the latest version after first registering their contact details at `http://www2.keil.com/mdk5/install/`.

How to do it...

1. Download the latest version of the software by following the instructions provided by Keil. Device-specific libraries are not included in installations from version 5 onwards, so at the end of the installation, we must configure the IDE using the Pack Installer to choose the resources (that is target devices, boards, and examples) that we need.

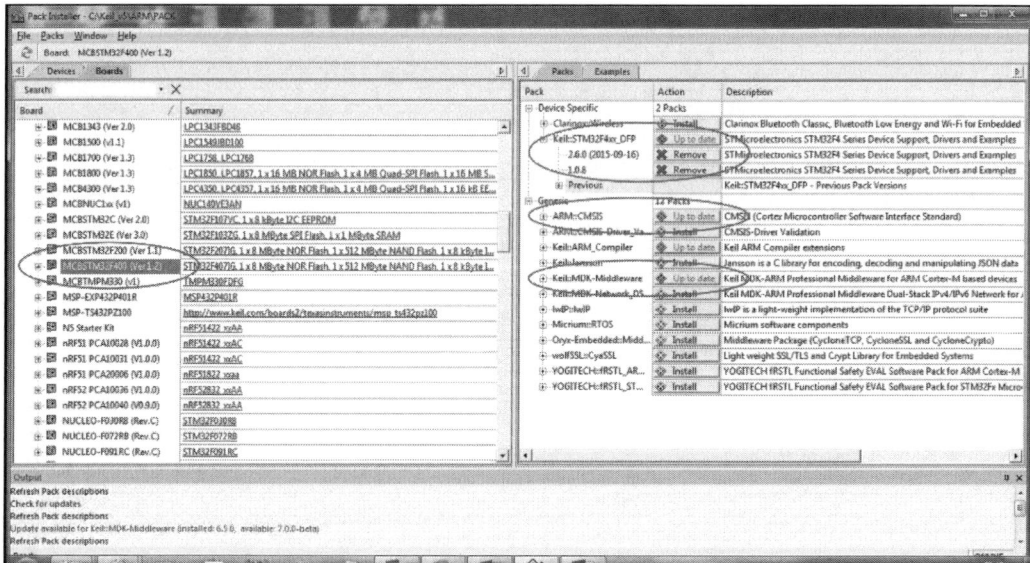

2. Select the **Boards** tab, choose the **MCBSTM32F400** Keil evaluation board featuring the STM32F407IGHx STMicroelectronics part, as this is the target for all the practical examples described in this cookbook.

3. With the **Packs** tab, in addition to the default installation options: CMSIS and Keil ARM Processional Middleware for ARM Cortex-M-based devices, board support for **MCBSTM32F400** is also needed. Select the latest version **Keil::32F4xx_DFP (2.6.0)**.

4. Select the **Examples** tab, and copy the board-specific example programs to a convenient local folder. Note: the example programs illustrate many useful features of the evaluation board, and are an invaluable resource.

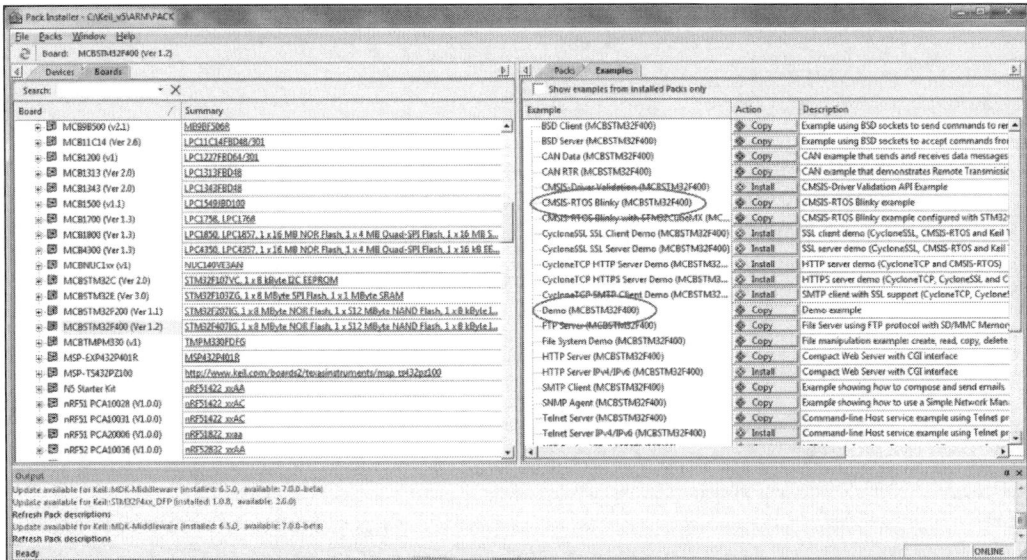

5. Once we have downloaded and installed MDK-ARM uVision5, the IDE can be invoked from the Windows Taskbar. If we wish to update the installation, the pack installer can be invoked by selecting the pack installer icon on uVision5 toolbar.

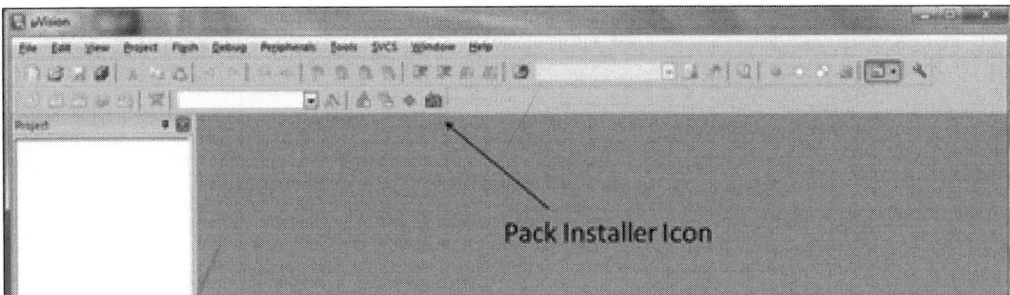

Pack Installer Icon

6. We demonstrate the basic features of uVision in this chapter, but later on, we'll probably need to access the uVision user guide via the Help menu (also available at `http://www2.keil.com/mdk5`) to learn about the more advanced features of the IDE. A useful guide to getting started with uVision5 can be found at `https://armkeil.blob.core.windows.net/product/mdk5-getting-started.pdf`. An overview of uVision5 is available at `http://www2.keil.com/mdk5`, and this includes some video clips that describe the design philosophy, and explain how to use the Pack Installer and create a new project.

How it works...

Computer programming involves specifying a sequence of binary codes that are interpreted by the machine as instructions that together enable it to undertake some task. The instruction sets of early computers were small and easily memorized by programmers, so programs were written directly in machine code, and each instruction code word was set up on switches and written to memory. Finally, once all the instructions had been entered, the program was executed. With the development of more powerful machines and larger instruction sets, this approach became unworkable. This motivated the need to program in higher level (human understandable) languages that are translated into machine code by a special program called a compiler. Modern day programmers rarely need to interpret individual binary codes; instead, they use a text editor to enter a sequence of high-level language statements, a compiler to convert them into machine code, a linker to allow programs to reuse previously written (library) code, and a loader to write the binary codes to memory. The steps comprising edit, compile, link, load can be undertaken by running each program (editor, compiler, linker, loader) separately. However, nowadays they are usually packaged together within a wrapper called an IDE. Some IDEs are language-specific and some are customizable, allowing developers to create bespoke programming environments for any target language and/or machine.

The pack installer framework allows MDK-ARM uVision5 to be customized and extended to target a large number of devices and evaluation boards using ARM cores. But while, IDEs represent the most popular and efficient route to programming, uVision represents just one of a number of IDEs that are widely available. Other manufacturers and open source communities offer alternatives, some of which we investigate later in the book.

Linking an evaluation board

This book focuses on the Keil STM32F400 evaluation board that features a STM32F407IGHx STMicroelectronics part to illustrate practical work. A wide range of other evaluation boards are available, and many of these are supported by the uVision5 IDE (that is, using the pack installer to download appropriate software components).

How to do it...

1. Once we have installed uVision, linking the evaluation board is simply a matter of connecting the two USB cables shown in the following image to your PC. The small daughter board shown in the image is Keil's **ULINK-ME** debug adaptor (`http://www.keil.com/ulinkme/`) that provides the data connection.

> The Windows plug-and-play feature will automatically find and install the driver (downloaded with uVision5).

2. The second USB cable provides power. Evaluation boards can usually be powered by a laptop or PC host connected via the USB port, but some laptop PSUs may be unable to supply sufficient current, and a USB hub might be required. Alternatively, an external supply can be connected via a separate power plug.

> The first time the ULINK device is used, its firmware needs to be configured. The configuration depends on the MDK version, and if we wish to use different versions of the MDK (that is, perhaps because we have legacy code developed using uVision4) then the ULINK configuration may need to be erased. `http://www.keil.com/support/docs/3632.htm` provides some further information and a download utility for this purpose.

How it works...

A USB-Link adaptor is needed to enable the executable code produced by the IDE to be uploaded to the evaluation board. The adaptor supports a **Joint Test Action Group** (**JTAG**) interface on the evaluation board, and offers a number of debugging possibilities (depending on the type of adaptor used). There are several debug adaptor connection options. Firstly, the Keil ULINK-ME debug adaptor (http://www.keil.com/ulinkme/), packaged together with the board as a starter kit, connects to the 20-pin JTAG connector and supports serial wire programming and on-chip debugging. Keil's ULINK-2 adaptor (http://www.keil.com/ulink2/) represents a more robust solution with similar functionality, and ULINK-Pro (http://www.keil.com/ulinkpro/) offers extended debug facilities employing high-speed streaming trace technology.

There's more...

The MCBSTM32F400 (http://www.keil.com/mcbstm32f400/) evaluation board shown in the preceding image features the STMicroelectronics STM32F407IGHx microcontroller part. The board specification includes the following:

- ▶ STM32F407IG Microcontroller
- ▶ On-chip and external memory
- ▶ 2.4 inch QVGA TFT LCD and touchscreen
- ▶ USB 2.0 Ports
- ▶ CAN interface
- ▶ Serial/UART Port
- ▶ Micro SD Card Interface
- ▶ 5-position Joystick
- ▶ 3-axis accelerometer
- ▶ 3-axis Gyroscope
- ▶ ADC Potentiometer input
- ▶ Audio Codec with Speaker and Microphone
- ▶ Digital Microphone
- ▶ Digital VGA Camera
- ▶ Push Buttons and LEDs directly connected to I/O ports
- ▶ Debug Interface

MCU manufacturers like Texas Instruments (TI), STMicroelectronics, Freescale, Atmel, Analog Devices, Silicon Labs, MikroElektronika, NXP, and Nordic Semiconductor all market evaluation boards featuring the Cortex-M4. Some of these offer cheaper, entry-level board options costing just a few dollars with functionality that can be enhanced by adding additional modules.

An insight into the range of microcontroller devices supported by MDK-ARM can be gained by scrolling through the list of packs listed by the Pack Installer. Keil markets a range of Cortex-M evaluation boards designed by themselves and other manufacturers (`http://www.keil.com/boards/cortexm.asp`) that feature a number of microcontrollers. Keil's range of boards features NXP, STMicroelectronics, and Freescale microcontrollers. The MCBSTM32 (Cortex-M3) and MCBSTM32F400 (Cortex-M4) evaluation boards offer one of the more expensive evaluation routes, but they are populated with a comprehensive set of I/O peripherals, including a QVGA TFT LCD touchscreen. STM (`http://www.st.com`) markets a similar evaluation board called the STM3241G-EVAL, offering almost identical features to Keil's but employing a slightly different PCB layout and using the STM32F417IG part.

Netduino (`http://netduino.com/`) offers a series of open source evaluation boards based on the STM32F405RG microcontroller featuring a Cortex-M4 core with open source software development support. Netduino is supported by an enthusiastic community of developers—a selection of projects which demonstrate the potential of the device are available.

Documentation for target devices and evaluation boards is available from the manufacturer. For example, those using the MCBSTM32F400 board will need to refer to the reference manual *RM0090* (`http://www.st.com`), the *MCBSTM32F200/400 User's Guide* (`http://www.keil.com`), the *ARM Cortex-M4 Processor Technical Reference Manual*, and the *Cortex-M4 Devices Generic User Guide* (`http://infocenter.arm.com`).

You will also find that the schematic diagram of the evaluation board, at `http://www.keil.com/mcbstm32f400/mcbstm32f400-schematics.pdf`, is also useful for resolving ambiguities in the libraries. If you use MDK-ARM, then once a new project has been created and the target microcontroller identified, most of the relevant documentation can be accessed via the **Books** tab within the project window.

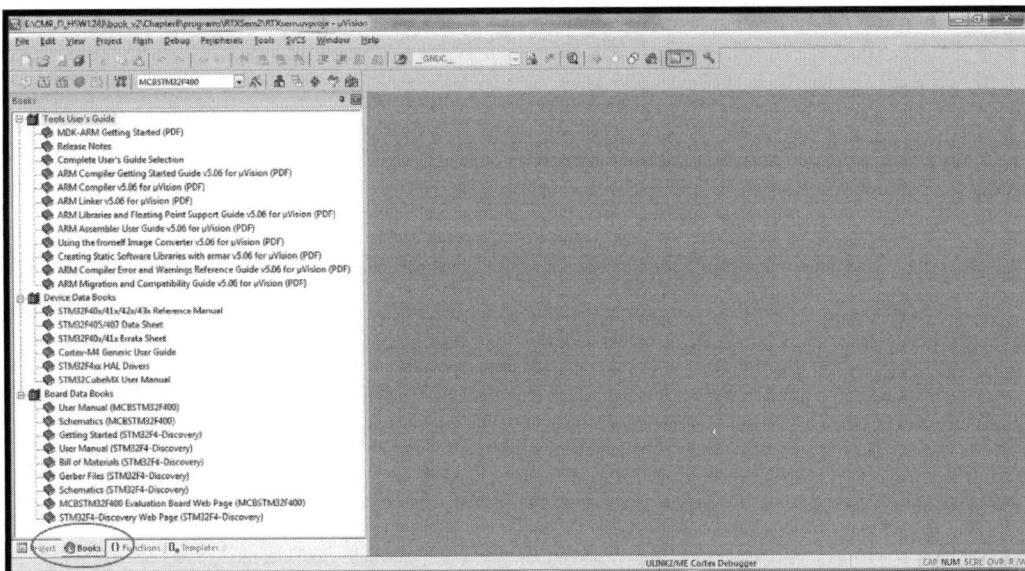

Running an example program

Manufacturers usually make a small number of example programs available that provide a tutorial introduction and demonstrate the potential of their evaluation boards. A simple program that flashes (that is, blinks) a **Light-emitting diode** (**LED**) on the board is usually provided. ANSI C is by far the most popular language amongst embedded system programmers, but other high level languages such as C++ and C# may also be supported. A brief introduction to the C programming language is provided in *Chapter 2, C Language Programming*.

The **Examples** tab in the pack installer for the STM32F4 series MCUs provides a link to a C program called **CMSIS-RTOS Blinky (MCBSTM32F400)** that flashes an LED connected to a GPIO port. The program is integrated within an MDK-ARM Project. Integrated development environments such as MDK-ARM usually manage software development tasks as projects, as in addition to the program source code itself, there are other target-specific details that are needed when the code is compiled. A project provides a good container for such things. We review the steps required to create a project from scratch in the next section.

How to do it...

1. Invoke uVision5. Open the Pack Installer, and copy the example program to a new folder (name the folder CMSIS-RTOS_Blinky).

2. Connect the evaluation board as described in the previous section. In addition to the ULINK cable, remember to connect a USB cable to supply power to your evaluation board.

3. Invoke uVision5 from the taskbar, select **Project → Open Project**; navigate to the folder named CMSIS-RTOS_Blinky, and open the file named blinky.uvprojx.

4. Build the project by selecting **Project → Rebuild all target files**, and then download the executable code to the board using **Flash → Download**. Take a moment to locate the **Build**, **Rebuild**, and **Download** shortcut icons on the toolbar as these save time.

5. Finally, press the **RESET** button on the evaluation board, and confirm that Blinky is running. You may notice that the Blinky example program does a little more than just flash one LED.

6. Once you have confirmed that your evaluation board is working, close the project (**Project → Close Project**), and quit uVision5.

How it works...

The program uses some advanced concepts such as CMSIS-RTOS (discussed in *Chapter 8, Real-Time Embedded Systems*.) to produce a visually interesting flashing LED pattern. We will not attempt to explain the code here, but the next section will develop a much simpler Blinky project called hello_blinky.uvprojx.

Writing a simple program

This section explains how to write, build, and execute a simple program. We also describe the various files that, together, make up a uVision project.

How to do it...

1. Use Windows Explorer to create a new (empty) folder called `helloBlinky_c1v0`. Invoke uVision5, and create a new project (**Project → New uVision Project...**). Navigate to the folder, and create a project file called `hello_blinky.uvprojx`. When prompted, choose the **STM32F407IGHx** device. Click **OK**.

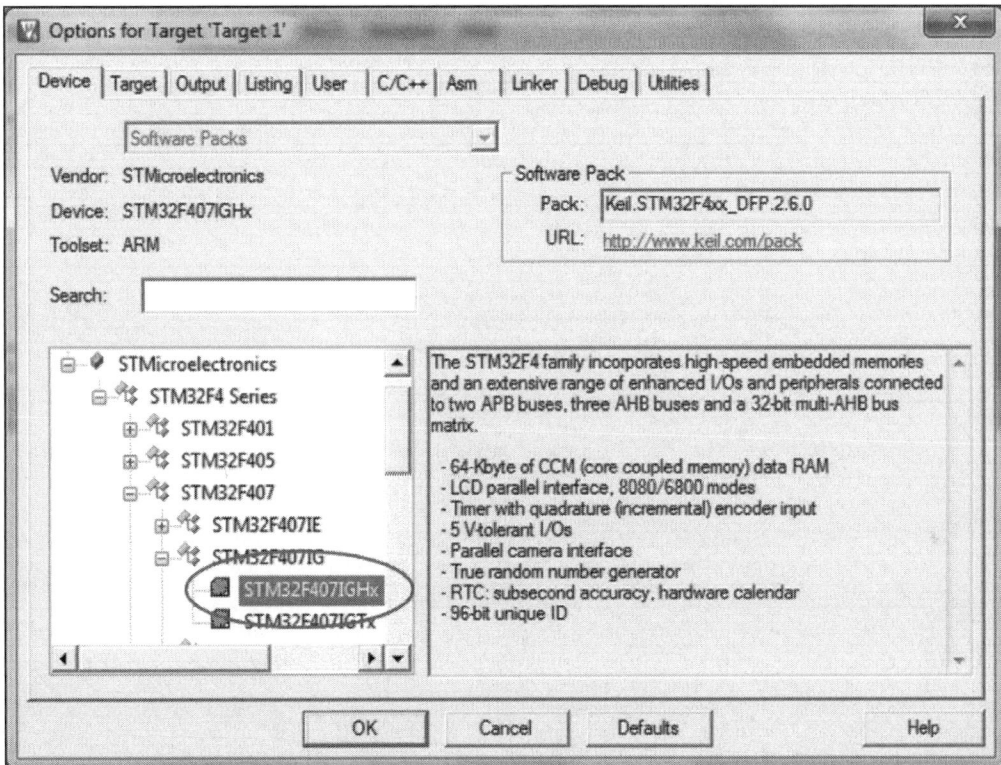

2. In **Manage → Run Time Environment**, choose the **MCB32F400** board support using the drop-down list, and tick the **LED** API (since our application will flash an LED). Expand the **Device** option list, and tick **Startup** and **Classic**.

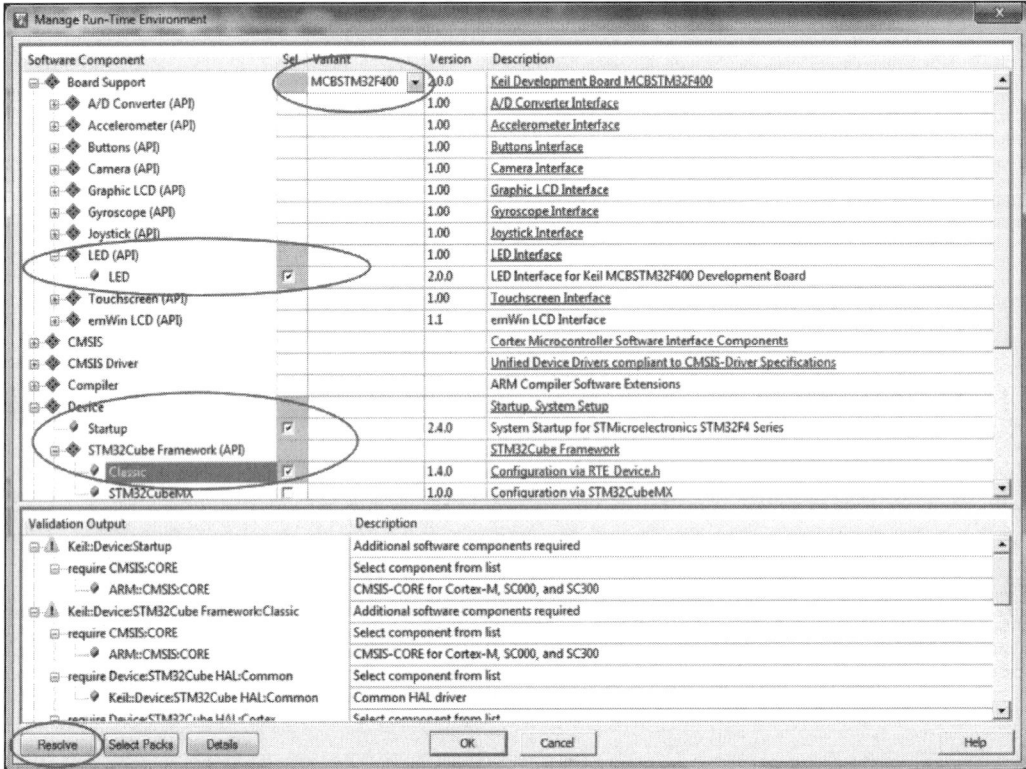

3. Notice that the **Validation Output** pane display warns us that, to drive LEDs, we also need CMSIS core, GPIO driver, and system start-up components. Press the **Resolve** button to automatically include any libraries needed by the board features selected, then click **OK**. The project window in uVision5 should show that the files have been successfully loaded. The names of the folders can be changed using a right-click menu, and fields can be expanded to show individual components, thereby allowing the file components to be edited. Note: Some library files are read-only.

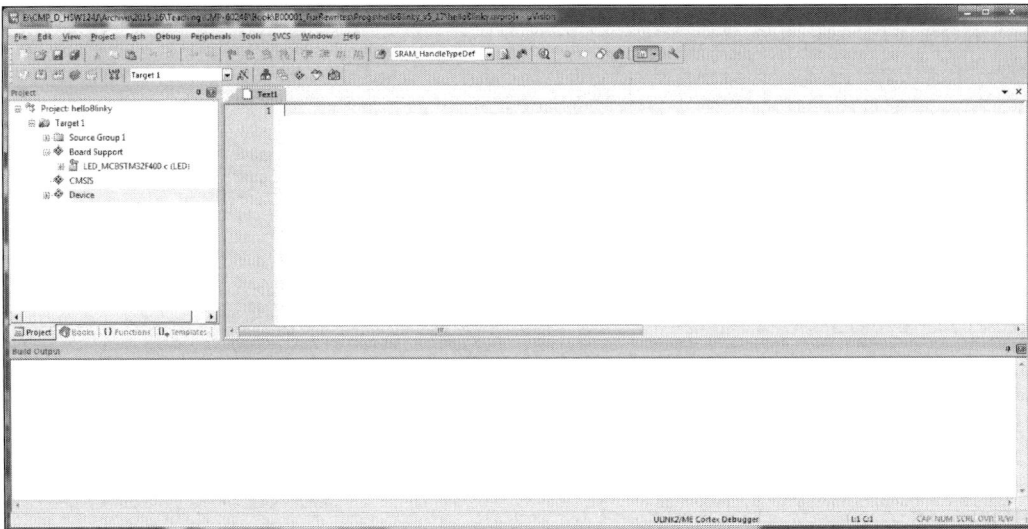

4. Right-click **Source Group 1**, and select **Add New Item to Group 'Source Group 1'...**; then select a C File (.c) template. Name the file `hello_Blinky.c`, and enter the following program:

```c
/*-------------------------------------------------
 * Recipe:   helloBlinky_c1v0
 * Name:     hello_blinky.c
 * Purpose: Very Simple MCBSTM32F400 LED Flasher
 *-------------------------------------------------
 *
 * Modification History
 * 16.01.14 Created
 * 27.11.15 Updated
 * (uVision5 v5.17STM32F4xx_DFP2.6.0)
 *
 * Dr Mark Fisher, CMP, UEA, Norwich, UK
 *-----------------------------------------------*/
#include "stm32f4xx_hal.h"
#include "Board_LED.h"

int main (void) {
  const unsigned int num = 0;
  unsigned int i;

  LED_Initialize();      /* LED Initialization */

  for (;;) {                          /* Loop forever */
```

```
                LED_On (num);            /* Turn specified LED on */
            for (i = 0; i < 10000000; i++)
                /* empty statement */ ;          /* Wait */
                LED_Off (num);          /* Turn specified LED off */
            for (i = 0; i < 10000000; i++)
                /* empty statement */ ;          /* Wait */
        } /* end for */
    }
```

5. The RTE manager of uVision5 will have configured the device options with values from the device database, but the debug options should be reviewed by selecting **Project → Options for Target 'MCBSTMF400'...** to ensure that they specify the **ULINK2/ME Cortex Debugger**.

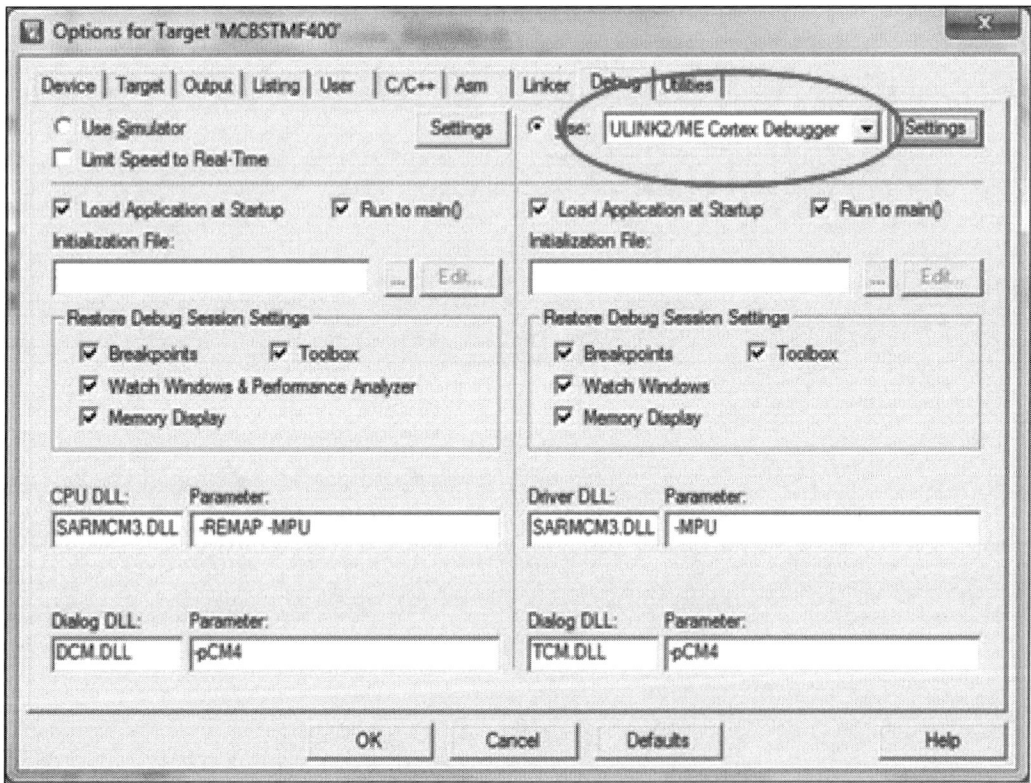

6. Build the project by selecting **Project → Rebuild all Target Files**. Again, there is a toolbar icon that provides a helpful shortcut.

7. Write the executable code to the microcontroller's flash memory using **Flash → Download**. Press the **RESET** button on the evaluation board to run the program.

> **Downloading the example code**
>
> You can download the example code files for all Packt books you have purchased from your account at http://www.packtpub.com. If you purchased this book elsewhere, you can visit http://www. packtpub.com/support and register to have the files e-mailed directly to you.

How it works...

Those familiar with uVision4 will notice that the most obvious feature for of this program is that a call to SystemInit() is missing, as this code is executed before main() is called. The function called main() is the entry point for our program, and each project should declare only one file that defines a main function. Conventionally, this might be called main.c, or adopt a file name that is shared by the project such as helloBlinky.c.

> Most of the file helloBlinky.c comprises comments, which are highlighted in green. Comments do not produce any executable code, but they are essential for understanding the program. You may be tempted to omit comments, but you will appreciate their value if, at some later date, you need to reuse code written by others, or even yourself.

The source code file begins with a large comment statement that extends over several lines and contains information about the program. Then there are C pre-processor directives; we discuss these in *Chapter 2, C Language Programming*. The program comprises a main function that declares two variables named i and num. There follows a function call to LED_Initialize() (written by developers) that sets up the GPIO peripheral which drives the LEDs. The program contains three so called for loops. The outer loop, is known as a superloop and never terminates. These statements within this loop are executed again and again, forever (well for as long as power is supplied to the evaluation board). The statements within the loop turn the specified LED ON and OFF by calling yet another function written by Keil developers. The other two for loops, nested within the superloop, simply waste time by incrementing the loop variable i. Implementing a delay in this way represents a very naïve approach, and we'll explore much more efficient techniques later. If you have not programmed in C before, then although you'll probably appreciate that this program is very compact, you may find it confusing. Don't worry, we'll revisit this program again when we introduce the C programming language in *Chapter 2, C Language Programming*.

There's more...

The structure of the uVision MDK projects has evolved considerably over the past few years and uVision5 represents a significant revision in this respect. Developers of uVision5 have attempted to make microcontroller software development much simpler by providing library functions that can be used to control peripherals such as LEDs, accelerometers, touchscreen, and so on. Many application developers migrating from uVision4 find this burdensome, and favor more classic approaches that do not rely on intrinsic interface functions. Application programmers who wish to use their own middleware functions are advised to download the ARMs MDK legacy support pack (`http://www2.keil.com/mdk5/legacy`). The source files that, together with the project options, define the `helloBlinky` project are summarized in the following table:

File Type	File extension	Description
C File	.c	Source code written in ANSI C.
Header File	.h	File containing additional information to be included in the source code
Assembly Language File	.s	Source code written in ARMs Thumb2 assembly language (Cortex-M cores)
Text File	.txt	Text file, usually containing description of the project or instructions for running the code.

A configuration wizard is provided to customize some files (for example, `startup_stm32F40xx.s`). However, we will deal with these more advanced aspects in subsequent chapters. Further, library and header file components, declared within the source files themselves, are also listed in the project window, and can be opened in the editor window. The file types you will encounter are described briefly in the following table, but will be discussed in more detail in *Chapter 2, C Language Programming*.

File Type	File extension	Description
C File	.c	Source code written in ANSI C.
Header File	.h	File containing additional information to be included in the source code
Assembly Language File	.s	Source code written in ARMs Thumb2 assembly language (Cortex-M cores)
Text File	.txt	Text file, usually containing description of the project or instructions for running the code.

The project options are functionally grouped together. They are accessed through the tabs within the **Project Options** menu, and summarized in the following table. Further details are available in the uVision User Guide.

Tab	Description
Device	Select the microcontroller device from the database
Target	Specify hardware parameters
Output	Define output files of the tool chain
Listing	Specify all listing files generated by the tool chain
User	Specify user programs executed before compilation / build
C/C++	Set C / C++ compiler-specific tool options
Asm	Set assembler-specific tool options such as macro processing
Linker	Set linker-related options, and define physical memory parameters.
Debug	Specify settings for the uVision debugger
Utilities	Configure utilities for flash programming

The options allow the developer to control quite small details of the build—for example, you might find it more convenient to execute code as soon as it is downloaded to the target by configuring the flash programming settings using the utilities tab as shown in the following image:

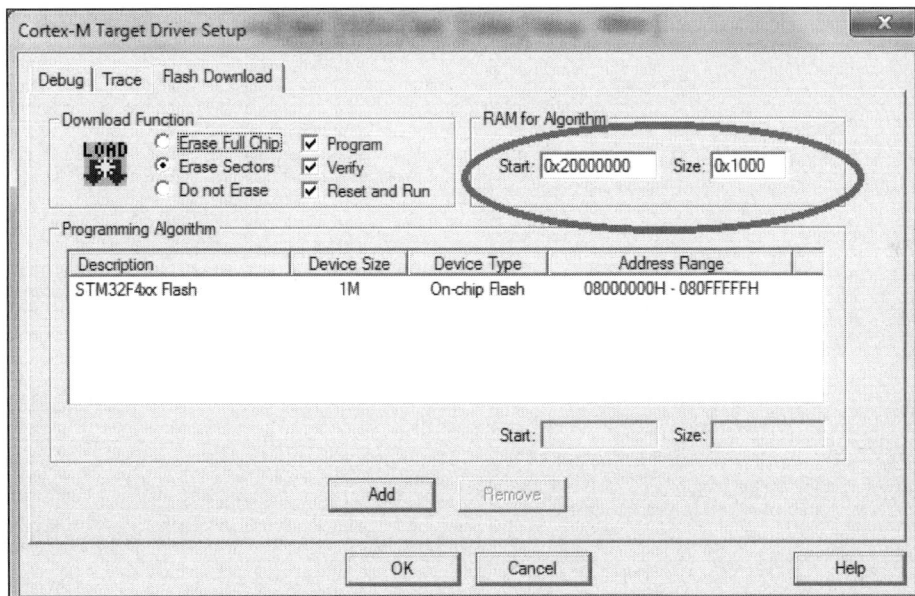

The STM32F400IGHx microcontroller implements 1MB On-chip Flash memory. **RAM for Algorithm** defines the address space used by the programming algorithm for the device.

Understanding the simple use of GPIO

Making an LED blink involves connecting it to a signal that alternately switches ON and OFF. **General purpose input/output (GPIO)** is the name of a microcontroller peripheral that provides functionality to source many signals at once (that is, in parallel). GPIO peripherals are designed to be very flexible, so configuring them can be rather confusing but using the RTE manager makes this process much simpler. We will modify our `helloBlinky_c1v0` recipe to simultaneously make all the LEDs blink rather than just one. Each LED on the evaluation board is connected to a pin on the microcontroller, so to illuminate an LED the microcontroller needs to provide a voltage and current similar to the that of a torch battery. To source this current, the corresponding GPIO port bit connected to the pin must be configured as an output that is switched ON and OFF by statements in our program that write to the port output data register.

How to do it...

To configure the GPIO follow the steps outlined:

1. Make a copy of the `helloBlinky_c1v0` folder from the previous recipe (and its contents) and rename this copy as `helloBlinky_c1v1`. Open the folder and open the `helloBlinky` project (double-click on the file). Then edit the `main` function defined in the `helloBlinky.c` file search for the following statement:

    ```
    LED_On (num);
    ```

2. Replace this statement with the following one:

    ```
    LED_SetOut (On_Code);
    ```

3. Also, search for the following statement:

    ```
    LED_Off (num);
    ```

4. Replace this statement with the following one:

    ```
    LED_SetOut (Off_Code);
    ```

5. The variables, `On_code` and `Off_Code`, are declared, as follows:

    ```
    const unsigned intOff_Code = 0x0000;
    const unsigned intOn_Code = 0x00FF;
    ```

6. A complete listing of the main function is as follows:

```
/*-----------------------------------------------------
 * Recipe:  helloBlinky_c1v1
 * Name:    helloBlinky.c
 * Purpose: Simultaneous MCBSTM32F400 LED Flasher
 *-----------------------------------------------------
 * Modification History
 * 16.01.14 Created
 * 03.12.15 Updated
 * (uVision5v5.17+STM32F4xx_DFP2.6.0)
 *
 * Dr Mark Fisher, CMP, UEA, Norwich, UK
 *-----------------------------------------------------*/
#include "stm32F4xx_hal.h"
#include "Board_LED.h"

int main (void) {
const unsigned intOff_Code = 0x0000;
const unsigned intOn_Code = 0x00FF;
   unsigned inti;

LED_Initialize();                         /* LED Init */

   for (;;) {                             /* Loop forever */
LED_SetOut (On_Code);             /* Turn LEDs on */
      for (i = 0; i< 1000000; i++)
/* empty statement */ ;                   /* Wait */
LED_SetOut (Off_Code);            /* Turn LEDs off */
      for (i = 0; i< 1000000; i++)
         /* empty statement */ ;                /* Wait */
   } /* end for */
}
```

7. Build, download, and run the application in exactly the same way as we did in the previous version.

How it works...

The GPIO interface is a particularly important feature in microcontrollers because it is designed to be easily integrated within user systems to drive light emitting diodes, read the state of switches, or connect to other peripheral interface circuits. Early I/O ports were prewired to provide either output or input interfaces, but soon they evolved into general purpose interfaces that could be programmed to provide either output or input connections. Later devices included more programmable features. As GPIO is so important for microcontroller applications, designers are keen to specify as many I/O pins as possible on their devices. However, increasing the device pin-out adds cost because the device becomes physically larger to accommodate the pins. This motivates manufacturers to develop devices that have pins that are configured by software. As you can imagine, configuring such a device is quite a challenge, so we're lucky that Keil's developers have provided library functions that make this task more manageable. As GPIO represents the interface between hardware and software, the evaluation board's schematic (`http://www.keil.com/mcbstm32f400/mcbstm32f400-schematics.pdf`) is essential to understanding the I/O.

The STM microcontroller used by the evaluation board provides eight GPIO ports, named A-I. Port pins PG6,7,8; PH2,3,6,7; PI10 are connected to LEDs. Those who have never encountered an LED may imagine it as a filament lamp, but an LED is a semiconductor device and behaves slightly differently. However, sticking with our initial lamp analogy (for the time being), we'll first consider a battery-operated torch comprising a battery, switch, and lamp. These components are connected by a copper wire that is often hidden within the body of the torch. We'll assume that the torch uses two AA batteries providing a voltage of about 3 Volts. We can depict the circuit as a diagram with symbols representing each of the components, as shown in the following diagram:

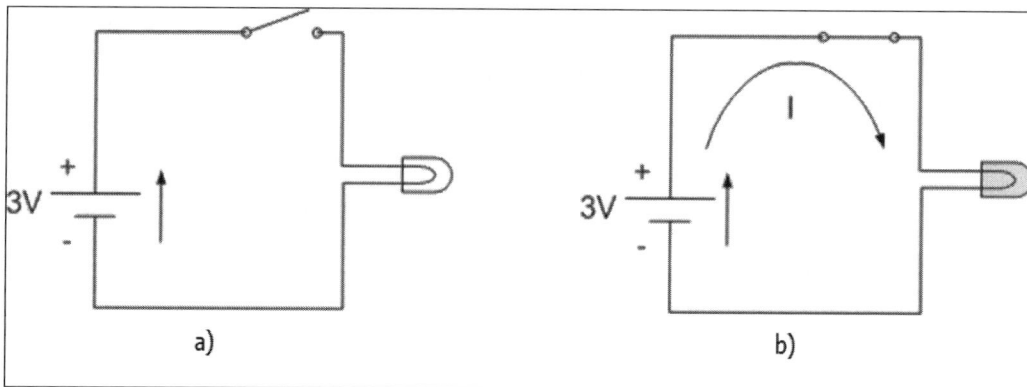

a) b)

When we close the switch, the battery voltage (denoted V) is applied directly to the lamp, a current flows (denoted I), heating the lamp filament, and this in turn, gives out light.

The electrical resistance (denoted R) of the filament determines the amount of current that flows according to Ohm's Law that is as follows:

$$I = \frac{V}{R}$$

Lamp filaments used in torches usually have a resistance of about 10 Ohms (10 Ω), so the amount of current flowing is about 0.3 A or 300 mA.

Imagine that a fault develops, which produces a short across the lamp. The current flowing is now only limited by the resistance of the copper wire and the internal resistance of the battery; these are both very small (a fraction of an Ohm). A high current will circulate which might, if the battery stored enough energy, cause the copper wire to heat up and melt the plastic case of the torch. However, AA batteries are unable to store sufficient energy for this to be a serious problem and in most cases the battery will discharge within a few seconds.

In modern torches, the lamp is replaced by an LED, which is a semiconductor device (its electrical properties lie between those of conductors, such as copper, and insulators, such as glass). An LED is a two terminal device with special properties. One of the terminals is known as the anode and the other as the cathode. If we replace the lamp in our torch with an LED, then current will only flow and the LED will illuminate when the anode is connected to the positive-battery terminal and the cathode to the negative-battery terminal, as depicted in the following diagram:

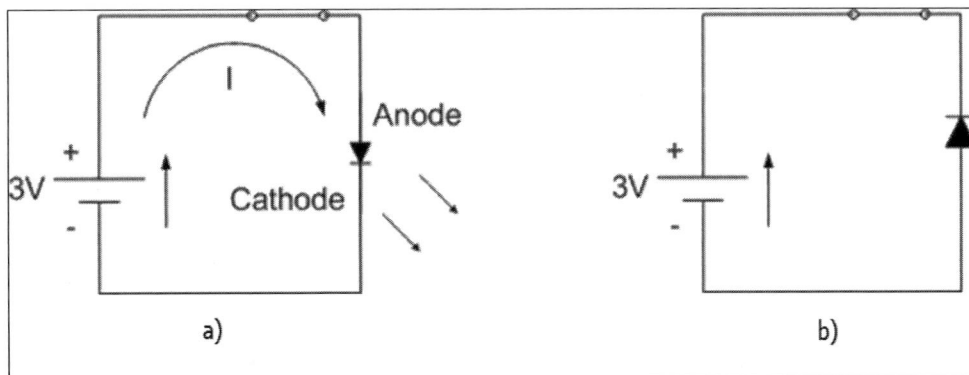

a)   b)

If we connect the device the other way round as depicted in the right side of the preceding diagram, then no current will flow; so, make sure that the batteries in your LED torch are fitted the right way round! When the anode is connected to the positive-battery terminal, the diode resistance is very low and the diode is said to be forward biased. When the cathode is connected to the positive-battery terminal the diode exhibits an extremely high resistance (negligible current flow) and the diode is said to be reversed biased. When forward biased, the LED exhibits an extremely low resistance, so an additional resistor must be placed in the circuit to limit the current flowing.

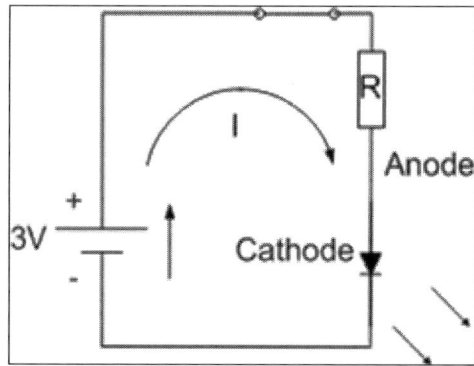

GPIO can also be used to read the state of switches that are connected to microcontroller pins. For this operation, each port bit must be configured as an input. When configured for input (that is, output is disabled), each bit of the parallel port's input data register is connected to a pin on the integrated circuit (on which the embedded processor is fabricated). Let's assume that we wish to connect a simple push-button switch to an input bit such that when the switch is operated, a voltage is applied to the port (pin), otherwise, no voltage is applied. The circuit a) shown as follows will achieve this. A complementary circuit that produces a voltage when the switch is open, and no voltage when the switch is operated (closed) is shown in b):

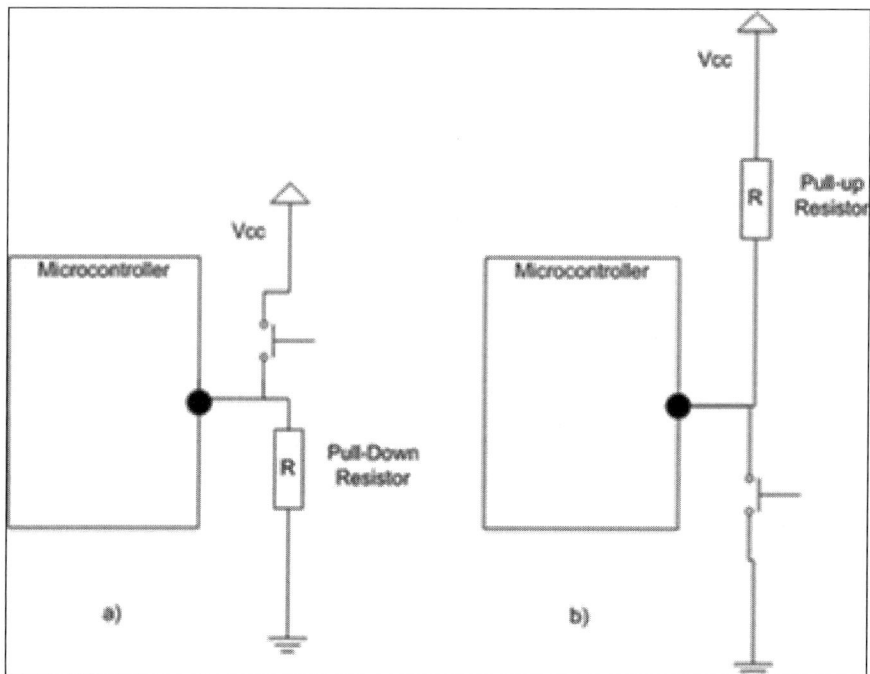

To eliminate the need for an additional resistor, the GPIO port input circuit includes one that can be configured by software as pull-up, pull-down, or disconnected. Obviously, when the port is configured as an output, both resistors are disconnected.

There's more...

Section 7 of STMicroelectronics Reference manual *RM0090* (www.st.com) for microcontrollers featuring the Cortex-M4 provides comprehensive programming details for the GPIO port. As well as producing logic signals (for example, making LEDs blink) and reading logic levels (for example, from switches), GPIO ports also provide an I/O path for other peripheral functions, such as Times and Digital-to-Analogue converters. We'll take a closer look at GPIO later on in this cookbook when we write programs that include more functionality.

Estimating microcontroller performance

The millions of instructions that can be executed per second (MIPS) is one measure of processor performance. This figure depends on the processor architecture, the clock speed, the memory performance, and so on. The microcontroller can be clocked from one of three oscillator sources. A **high speed external** (**HSE**) clock is derived from a 25 MHz crystal oscillator connected between two pins of the microcontroller. A **high speed internal** (**HSI**) clock is sourced from an internal 16 MHz **resistor-capacitor** (**RC**) controlled oscillator, and a **Phase Locked Loop** (**PLL**) can be configured to provide multiples of either HSE or HSI.

A peripheral called **reset and clock control** (**RCC**) allows the clock source to be selected and configured using a circuit known as a clock tree. The RCC peripheral also sources clocks for other microcontroller peripherals, and these also need to be configured. Following a hard reset, the RCC configuration is determined by the RCC register default values given in the *RM0090* Reference Manual (www.st.com). Selecting **Startup** from the **Device** submenu of the RTE manager copies an assembly language file named startup_stm32f407xx.s (the .s file extension is conventionally used to identify assembly language files) to our project. This file holds the exception table. The reset exception generated by a hard reset (that is, activating the reset button on the evaluation board) causes the microcontroller's program counter to be loaded with the address of the reset handler (identified by symbol Reset_Handler), and this in turn calls a function named SystemInit() defined in the file, system_stm32f4xx.c. This function configures the RCC to use the 16 MHz HSI clock before calling the function main().

How to do it...

1. Run `helloBlinky`, and measure the frequency of the 'blinks'. We should see about 4 blinks/second or 4 Hz. It may be easier to count the blinks in a 10-second period.

2. When we examine the program code shown earlier, we see that the program spends most of its time executing the two nested `for` loops. The statements inside these loops are executed thousands of times. Some readers may have spotted that there are no statements called inside the loop; but even so, the loop counter must be updated on each iteration. This operation requires a addition (ADD) instruction followed by a compare (CMP) instruction to be executed.

3. We need to do some elementary math to work out how long it will take to execute these two instructions. Checking Table 3.1 of the *ARM Cortex-M4 Processor Technical Reference Manual*, we see that these each take 1 cycle to execute. Since `SystemInit()` configures the RCC to use the HSI (16 MHz)clock, the time needed to switch the LED ON/OFF once will be *2 X (1000000) x 1/(16 x 106) x 2 (instructions) = 250 ms* (that is, about 4 times per second).

There's more...

To understand how the processor achieves this level of performance, we need to look at the processor architecture. The processor implements the ARMv7-M architecture profile described at http://infocenter.arm.com. ARMv7-M is a 32-bit architecture and the internal registers and data path are all 32-bit wide. ARMv7-M supports the Thumb Instruction Set Architecture (ISA) with Thumb-2 technology that includes both 16 and 32-bit instructions. ARM processors were originally inspired by **Reduced Instruction Set Computing** (**RISC**) architectures developed in the 1980s. RISC architecture attempted to improve on the performance of traditional computer architectures of the era that employed the so-called **Complex Instruction Set Computing** (**CISC**) architectures, by defining an ISA that supported a small number of instructions, each of which could be executed in one processor clock cycle, and so achieve a performance advantage. In the three decades since RISC was proposed, the size and complexity of RISC ISA's has increased, but the goal is still to minimize the number of clock cycles needed to execute each instruction. With this in mind, ARM Cortex-M3 and M4 processors have a three-stage instruction pipeline and Harvard bus architecture. Computers that use Harvard architecture have separate memories and busses for instructions and data rather than the shared memory systems used by von Neumann architectures, and the higher memory bandwidth this affords can achieve better performance.

The Cortex-M4 processor also provides signal processing support including a **Single Instruction Multiple Data** (**SIMD**) array processor and a fast **Multiply Accumulator** (**MAC**). Together with an optional **Floating Point Unit** (**FPU**), these features allow the Cortex-M4 to achieve much higher performance in **Digital Signal Processing** (**DSP**) applications than the earlier Cortex-M3.

See also

Besides manufacturers' data sheets, there are a few books that address the Cortex-M4. Joseph Yiu's books (`http://store.elsevier.com/Newnes/IMP_73/`) on the Cortex-M3 and M4 processors are aimed at programmers, embedded product designers, and System-on-Chip (SoC) engineers. Books for undergraduate courses include a series of books by Jonathan Valvano (`http://users.ece.utexas.edu/~valvano`) and a text written by Daniel Lewis (`http://catalogue.pearsoned.co.uk`). Trevor Martin has also written an excellent guide to STM32 microcontrollers. This document is one of a number of insider guides that can be downloaded from `http://www.hitex.com`.

2

C Language Programming

In this chapter, we will cover the following topics:

- ▶ Configuring the hardware abstraction layer
- ▶ Writing a C program to blink each LED in turn
- ▶ Writing a function
- ▶ Writing to the console window
- ▶ Writing to the GLCD
- ▶ Creating a game application – Stage 1
- ▶ Creating a game application – Stage 2
- ▶ Debugging your code using print statements
- ▶ Using the debugger

Introduction

This chapter will introduce you to writing programs in C, a high-level language developed in the 1970s and popular amongst embedded system developers. It is not the only high-level language that can be used to target embedded system applications, but it is the most widely used, because it produces executable code that is compact and very efficiently executed. Standards for C are published by the American National Standards Institute (ANSI) and the International Organization for Standardization (ISO). The current standard for the C Programming Language (C11) is ISO/IEC 9899:2011 (`http://www.open-std.org/jtc1/sc22/wg14/www/standards`).

Becoming a competent C programmer will take time, and although this chapter provides a starting point, you will undoubtedly need to consult other texts that provide a more thorough treatment of the topic. There are also a number of online resources such as http://crasseux.com/books/ctutorial/ and http://www.csd.uwo.ca/~jamie/C/index.html.

Configuring the hardware abstraction layer

The method we deployed in *Chapter 1, A Practical Introduction to ARM® CORTEX®* used Startup.c to provide a very basic **Run Time Environment** (**RTE**), and although this is sufficient to get started blinking LEDs, we need to define a more advanced RTE to take advantage of the other peripherals we'll meet in future recipes. The Application Programmers Interface (API) that STMicroelectronics (STMicro) provide for their microcontrollers is called a **hardware abstraction layer** (**HAL**), and CMSIS v2.0 compliant programs must configure this before initializing their peripherals. The RTE manager offers two routes named Classic and STM32CubeMX to configure the HAL. Selecting STM32CubeMX invokes a graphical tool developed by STMicro (freely available at www.st.com) that creates the RTE (that is, generates RTE.h and imports the associated libraries). We describe this process in *Chapter 9, Embedded Toolchain*. Since we're already familiar with the Classic API, we'll continue to use this, and simply add a few lines of code to configure the HAL.

How to do it...

For configuring the HAL follow the steps outlined:

1. Make a copy of the folder helloBlinky_c1v1 which we created in *Chapter 1, A Practical Introduction to ARM® CORTEX®, Understanding the simple use of GPIO* and name it helloBlinky_c2v0.

> Copying a folder and renaming it is a quick way to extend an existing project. Future recipes refer to this process as cloning the project.

2. Open the project, and using the RTE manager, expand the **CMSIS→RTOS (API)** software component. Check the **KeilRTX** option. Click on **Resolve**, and exit using **OK**.

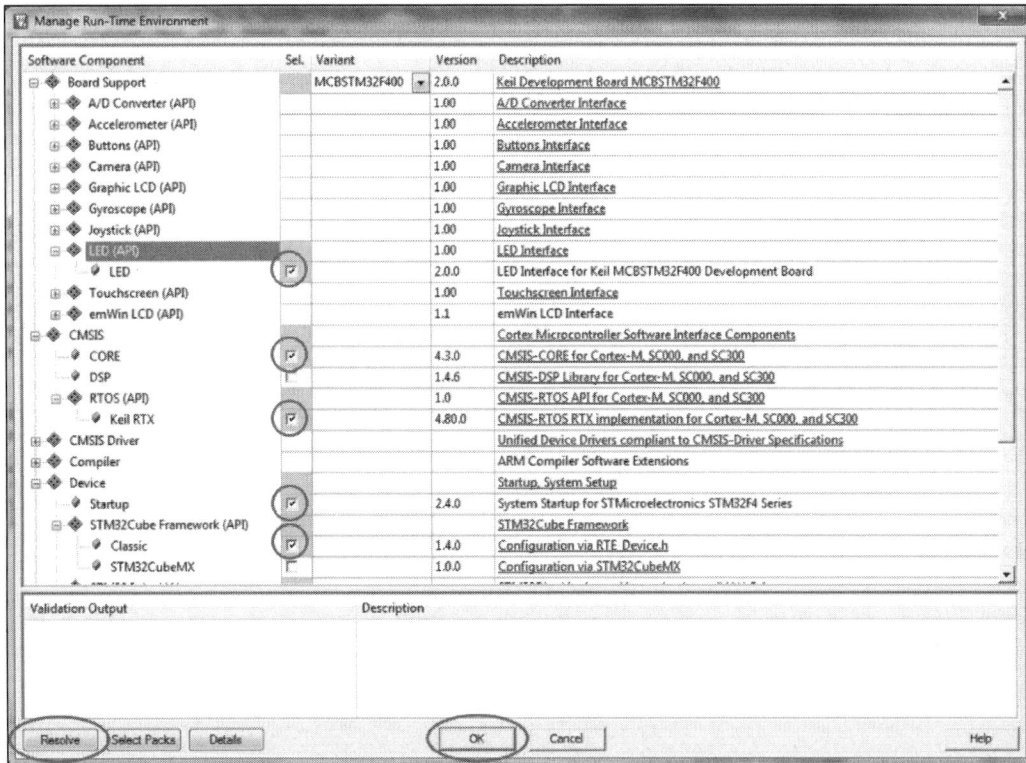

3. Add #include "cmsis_os.h"

4. Add a function prototype declaration, that is, void SystemClock_Config(void) in the file helloBlinky.c.

5. Add the following lines of code (copy and paste from the example project CMSIS-RTOS Blinky):

```
#ifdef __RTX
extern uint32_t os_time;

uint32_t HAL_GetTick(void) {
  return os_time;
}
#endif
```

6. Copy the function named `SystemClock_Config ( )` from the example project `CMSIS-RTOS Blinky`, and paste this into the file `helloBlinky.c`.

7. Add calls to `HAL_Init ( )` and `SystemClock_Config ( )` at the beginning of `main()`. Our source code file `helloBlinky.c` should now appear as follows:

```c
#include "stm32f4xx_hal.h"
#include "Board_LED.h"
#include "cmsis_os.h"

/* Function Prototype */
void SystemClock_Config(void);

#ifdef __RTX
extern uint32_t os_time;

uint32_t HAL_GetTick(void) {
  return os_time;
}
#endif

/**
  * System Clock Configuration
  */
void SystemClock_Config(void) {
  RCC_OscInitTypeDef RCC_OscInitStruct;
  RCC_ClkInitTypeDef RCC_ClkInitStruct;

  /* Enable Power Control clock */
  __HAL_RCC_PWR_CLK_ENABLE();

  /* The voltage scaling allows optimizing the power
     consumption when the device is clocked below the
     maximum system frequency (see datasheet). */

  __HAL_PWR_VOLTAGESCALING_CONFIG
     (PWR_REGULATOR_VOLTAGE_SCALE1);

  /* Enable HSE Oscillator and activate PLL
      with HSE as source */
  RCC_OscInitStruct.OscillatorType =
    RCC_OSCILLATORTYPE_HSE;
  RCC_OscInitStruct.HSEState = RCC_HSE_ON;
  RCC_OscInitStruct.PLL.PLLState = RCC_PLL_ON;
  RCC_OscInitStruct.PLL.PLLSource = RCC_PLLSOURCE_HSE;
  RCC_OscInitStruct.PLL.PLLM = 25;
```

```
    RCC_OscInitStruct.PLL.PLLN = 336;
    RCC_OscInitStruct.PLL.PLLP = RCC_PLLP_DIV2;
    RCC_OscInitStruct.PLL.PLLQ = 7;
    HAL_RCC_OscConfig(&RCC_OscInitStruct);

    /* Select PLL as system clock source and configure
       the HCLK, PCLK1 and PCLK2 clocks dividers */
    RCC_ClkInitStruct.ClockType = RCC_CLOCKTYPE_SYSCLK |
                                  RCC_CLOCKTYPE_PCLK1 |
                                  RCC_CLOCKTYPE_PCLK2;
    RCC_ClkInitStruct.SYSCLKSource =
                            RCC_SYSCLKSOURCE_PLLCLK;
    RCC_ClkInitStruct.AHBCLKDivider = RCC_SYSCLK_DIV1;
    RCC_ClkInitStruct.APB1CLKDivider = RCC_HCLK_DIV4;
    RCC_ClkInitStruct.APB2CLKDivider = RCC_HCLK_DIV2;
    HAL_RCC_ClockConfig(&RCC_ClkInitStruct,
                                FLASH_LATENCY_5);
}

/**
  * Main function
  */
int main (void) {
  const unsigned int Off_Code = 0x0000;
  const unsigned int On_Code = 0x00FF;
  unsigned int i;

  HAL_Init ( );   /* Init Hardware Abstraction Layer */
  SystemClock_Config ( );        /* Config Clocks */
  LED_Initialize ( );              /* LED Init */
  // etc...
}
```

8. Build and run the program.

> Notice that the code executes about 10 times faster than the recipe of *Chapter 1, A Practical Introduction to ARM® CORTEX®*. Try commenting out the call `SystemClock_Config ( )` in `main ( )` by placing `//` immediately before the statement. Rebuild and run. Compare the execution speed of the two versions.

How it works...

The function `SystemClock_Config ( )` comprehensively configures the clock tree shown in *Figure 16* of STMicro's reference manual *RM0090* (www.st.com). It selects the **Phase Locked Loop** (**PLL**) clock derived from the 25 MHz crystal controlled HSE clock as the System Clock, and configures the multiplier *N = 336* and dividers *P = 2* and *M = 25*. The system clock frequency is given by:

$$\text{Tx/Rx baud} = \frac{f_{clk}}{8(2 \times \text{OVERS}) \times \text{USARTDIV}}$$

The configuration values are held in two data structures (structs) called `RCC_OscInitStruct` and `RCC_ClkInitStruct`.

As we will see later in the chapter, functions may be declared implicitly by the function definition or explicitly by a function prototype. Function prototypes are considered to be preferable, and these are often declared in header files (for example, see `Board_LED.h`). So, in case we've given a prototype declaration first,

Structs just identify the arrangements of data in memory. We will discuss structs later once we've dealt with more basic data types such as integers.

Finally, the following section of code:

```
#include "cmsis_os.h"

#ifdef __RTX
extern uint32_t os_time;

uint32_t HAL_GetTick(void) {
   return os_time;
}
#endif
```

It isn't strictly necessary for a program that only uses GPIO, but subsequent recipes using other peripherals need it. So, to avoid illustrating the configuration each time, we'll assume this boilerplate is included in all future recipes.

Lastly, we've called our source code file `helloBlinky.c`. This is the same name we gave the project. By convention, this indicates that this source code file contains the `main()` function.

Writing a C program to blink each LED in turn

This recipe extends the `helloBlinky_c2v0` recipe introduced in the previous section, and includes a few more C programming statements. We'll call our new recipe `helloBlinky_c2v1`. uVision5's IDE features a so-called folding editor that allows blocks of code and comments to be hidden or expanded. This is quite useful for hiding complexity, allowing us to focus on the important details.

Getting ready...

First, we'll draw a flowchart describing what our program will do. Don't worry about the details at this stage, we just need to describe the behavior. A flowchart describing `helloBlinky_c2v1` is shown as follows:

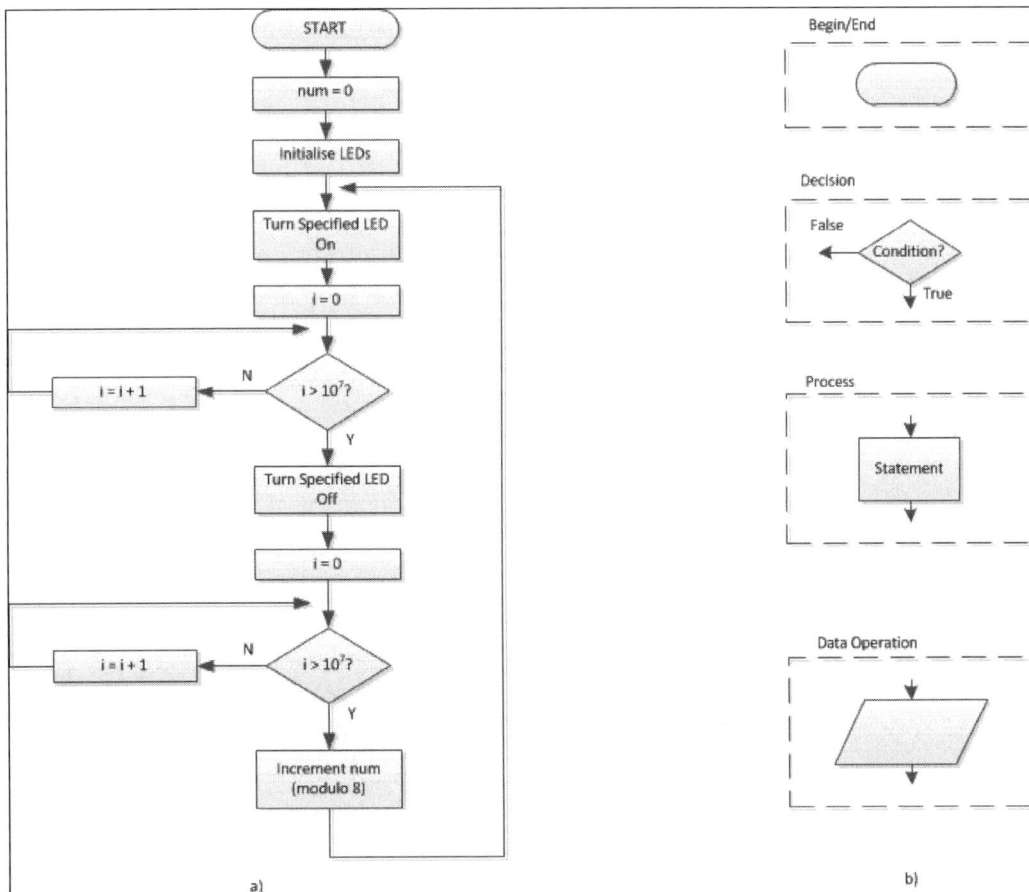

Our program will need to change the value of a number stored in memory that determines the LED that is illuminated. Numbers coded in this way are called variables. The name of the variable is chosen by the programmer (usually programmers try to pick meaningful names); in this case, it's referred to by the identifier `num`. Since there are only eight LEDs, the values we assign to `num` are 0,1,2,3,4,5,6, and 7. The subroutines `LED_On` and `LED_Off` use the variable to determine which LED is switched.

The flowchart illustrates several different types of operation, identified by the geometrical shapes shown in the preceding diagram as follows:

- **Diamond**: A decision operation with two outcomes Yes (True) or No (False)
- **Rectangle**: A process operation
- **Parallelogram**: A data operation
- **Rounded rectangle**: Start/End

Within the flowchart, we can identify processes that are executed within a loop, and so are repeated until a certain condition is fulfilled. Structures such as this are a common feature in algorithms, and high-level programming languages have evolved to enable such operations to be efficiently coded.

How to do it...

1. Clone `helloBlinky_c2v0` to create `helloBlinky_c2v1`.
2. Modify `main()` as follows (keep the boilerplate unchanged):

```c
int main (void) {
  unsigned int i;
  unsigned int num;

HAL_Init ( );                /* Init Hardware Abstraction Layer */
SystemClock_Config ( );      /* Config Clocks */
LED_Initialize ( );          /* LED Init */

  for (;;) {                          /* Loop forever */
    LED_On (num);                     /* Turn LEDs on */
    for (i = 0; i < 1000000; i++)
     /* empty statement */ ;              /* Wait */
    LED_Off (num);              /* Turn LEDs off */
    for (i = 0; i < 1000000; i++)
      /* empty statement */ ;             /* Wait */
    num = (num+1)%8;    /* increment num (modulo-8) */
  } /* end for */
}
```

3. Once we have entered the code, we build it and download it to the evaluation board in exactly manner as we did for the `helloBlinky_c2v0` recipe.

4. Run the program by pressing **RESET** on the evaluation board.

How it works...

The program starts with two statements beginning with a # character. These are not program statements but directives for the C preprocessor. The preprocessor resolves all these directives before the C compiler parses the rest of the code. It is considered good practice to group these together at the start of the program. Preprocessor directives can only extend over one line, and they are not terminated by a semicolon. However, to aid readability, longer directives can be split over several lines by using a \ character to terminate each block of text. There are six types of directives:

- ▸ **Macro definition**: `#define` and `#undef`
- ▸ **Conditional inclusion**: `#ifdef`, `#ifndef`, `#if`, `#endif`, `#else`, and `#elsif`
- ▸ **Line control directive**: `#line`
- ▸ **Error directive**: `#error`
- ▸ **File inclusion**: `#include`
- ▸ **Pragma directive**: `#pragma`

We'll briefly explain these directives as they are introduced in the recipes we consider. However, there are plenty of online resources available for those who feel they need more detail (for example, `http://gcc.gnu.org/onlinedocs/cpp/`). The preprocessor parses the headers:

```
#include "stm32f4xx_hal.h"
#include "Board_LED.h"
#include "cmsis_os.h"
```

replacing each `#include` directive with the contents of the files `stm32f4xx_hal.h`, `Board_LED.h` and `cmsis_os.h`. By convention, include files adopt `.h` file extensions, while those not included in other files are given a `.c` file extension. Later on, we'll meet another style of `#include` directive:

```
#include <stdio.h>
```

In this case, the filename is enclosed in angled brackets. This syntax is used to indicate that the compiler's standard include path is to be searched. When the filename is enclosed in double quotes, the search path includes the current directory. We can add folders in the include path, and select compiler options using the C/C++ tab in the project options window.

The next statement declares a function called `main()`. Every C program must include one (but only one) function named `main()`. The structure of the `main()` function of all the embedded C programs that we'll meet is as follows:

```
int main (void) {

    . . .

}
```

We identify the input arguments (args) of `main()` inside the brackets; in this case, there are none, and so we use the reserved word `void` to indicate none are to be expected. Before `main()` we see (primitive data type) `int`, indicating that `main()` returns an integer. Conventionally, `main()` returns a value 0 to indicate to the program that called `main()` (that is, the operating system) that the program terminated successfully. But since our program doesn't run under an operating system and typically declares an infinite loop (called a superloop), there is no need to include a return statement at the end of `main()` (if we do, the compiler will warn us that it's not reachable). The other feature of `main()` are the braces, { and }, that are used to identify the beginning and end of the block of statements that comprise `main()`. Note that the curly bracket (opening brace) immediately following `main()` is paired with the closing brace that terminates the statements within `main()`. These braces mark the beginning and end of the `main()` function; the statements inside the braces belong to `main()`. We indent these statements to make this clearer. The first two statements in `main()` are variable declarations. Because C is a strongly-typed language, we must declare all our variables before we use them. In so doing, we're telling the compiler how many bits to use to represent the number so that it can determine the size of the memory space needed to store them.

The values that a computer manipulates are stored in binary. In the binary system, number values are represented by a sequence of digits, just like the decimal system. However, whereas the decimal system uses digits 0,1,2,3,4,5,6,7,8, and 9, the binary system uses only 0 and 1. Digits 0 and 1 in the binary number system are called **bits**.

The decimal system is a positional number system, where the value of the number is determined by the position of the digits relative to the decimal point. Conventionally, when we write whole numbers, we assume the decimal point is immediately to the right of the least significant digit. Hence, if there are three digits, each represents (from left to right) the number of hundreds (10^2), tens (10^1), and units (10^0), for example:

$365_{10} = (3 \times 10^2) + (6 \times 10^1) + (5 \times 10^0)$

Consider a similar 3-bit binary number. Here, each bit represents (from left to right) multiples of 2^2, 2^1, and 2^0, for example:

$101_2 = (1 \times 2^2) + (0 \times 2^1 + (1 \times 2^0) = 5_{10}$

In the preceding examples, we are using a subscript to represent the base (or radix) of the number system just to avoid any confusion.

Inside a computer, each bit is represented as an electrical signal; typically a +ve signal voltage represents a '1' and no voltage (0 v) represents '0'. To manipulate a 3-bit binary number, a computer must provide three signal transmission paths, and the registers within the Central Processing Unit (CPU) must be capable of storing 3 bits. You have probably already spotted that three bits isn't going to be of much use, as a 3-bit computer can only manipulate quantities between 0_{10} and 7_{10}. Historically, some simple 3-bit computers have been used for elementary control tasks, but many more have been designed to manipulate 8, 16, 32, and 64 bits. The number of bits that a computer has been designed to manipulate is called its word length. As we've seen, the ARM Cortex has been designed with 32-bit registers (that is, a 32-bit word length). A typical ARM Cortex register can be visualized as 32 cells, each able to store 1 bit of data:

2^{31}																															2^0
1	1	0	0	1	0	0	0	0	0	1	0	0	0	1	0	0	0	0	0	0	0	0	0	0	0	0	0	1	0	0	1
MSB																															LSB

The preceding register is shown storing a binary representation of the decimal number:

$$(1 \times 2^{31}) + (1 \times 2^{30}) + (1 \times 2^{27}) + (1 \times 2^{21}) + (1 \times 2^{17}) + (1 \times 2^{3}) + (1 \times 2^{0}) = 3357671363_{10}$$

A 32-bit register can store positive numbers between 0 and $(2^{32}-1)$, that is, $(0 - 4294967199_{10})$. Most of us (me included!) need a pocket calculator to convert between binary and decimal (and vice versa), so we need a more human-friendly way of efficiently representing binary quantities. Hexadecimal (radix 16) representations provides this by allowing groups of 4 bits (representing $0\text{-}15_{10}$) to be mapped to digits 0,1,2,3,4,5,6,7,8,9,A,B,C,D,E, and F , that is:

1	1	0	0	1	0	0	0	0	0	1	0	0	0	1	0	0	0	0	0	0	0	0	0	0	0	0	0	1	0	0	1
C				8				2				2				0				0				0				9			

Hence, $1100100000100010000000000001001_2 = 3357671363_{10} = C8220009_{16}$. We identify hexadecimal (hex) numbers in C programs using the syntax 0xC8220009. In this case, since there are 8 hex digits, we have an 8 x 4 = 32-bit binary word.

The number of bits used to represent a number is determined by its data type. Some of the more common basic (also called primitive) C data types are:

- char (8-bit)
- short int (16 bits)
- unsigned short int (16 bits)
- int (32 bits)
- unsigned int (32 bits)
- long int (64 bits)
- unsigned long int (64 bits)

A full list of basic types is available at `https://en.wikipedia.org/wiki/C_data_types`. Data types qualified by the identifier `unsigned` indicate that the value should be interpreted as representing only positive quantities. Sometimes, embedded developers define aliases for the basic data types, such as `int32_t`, `uint32_t`, and so on. We'll explain the purpose of this in *Chapter 3, Assembly Language Programming* but for the time being, don't be concerned if you see these identifiers used in library functions.

The `helloBlinky_c1v1` recipe of *Chapter 1, A Practical Introduction to ARM® CORTEX®* declares two variables, both 32 bits in length:

```
const unsigned int num = 0;
unsigned int i;
```

The first variable declaration is preceded by the qualifier `const` and assigned a value 0. The `const` qualifier tells the compiler to treat the variable as a constant, and so, if we attempt to change its value in a subsequent assignment statement, then the compiler will issue an error. When a variable is declared, the compiler just reserves somewhere to store it; this might be in a register (registers are places that data can be stored in the processor) or in memory. Values are assigned to variables by assignment statements; for example,

```
p = 0;
```

places 0 in the memory location or register referenced by the identifier `p`.

To generate a more interesting LED lightshow, we'll need to write to a different LED each time we execute the superloop. We use the functions `LED_On()` and `LED_Off()` to switch the LEDs (as we did in `helloBlinky_c1v1`), but this time, we increment that value of the variable (`num`) that controls the LED that we switch each time we iterate the superloop. Since there are 8 LEDs (`num = 0` represents the Least Significant LED and `num = 7` the Most Significant), we need `num` to behave as a modulo-8 counter (that is, 7+1 = 0). The statement

```
num = (num+1)%8;
```

achieves this. The `%` operator performs modulo division. Of course, we don't need the `const` qualifier in the declaration for `num`, as its value is changed within `main()`. Variable `i` is used by the for loop to implement a delay in exactly the same way as it was in our `helloBlinky_c1v1` recipe.

There's more...

High-level languages such as C typically provide mechanisms that allow the programmer to express decisions and iterations within the algorithm by means of IF, FOR, and WHILE structures shown in the following diagram (a). uVision5 provides common templates shown in (b) to help the programmer include these structures in their code.

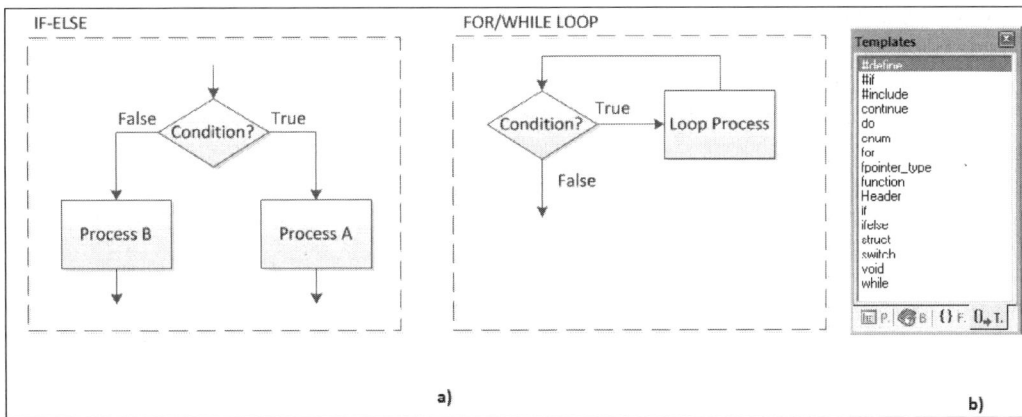

IF-ELSE

False ◇ Condition? True

Process B   Process A

FOR/WHILE LOOP

◇ Condition? — True → Loop Process

False

Templates
#define
#if
#include
continue
do
enum
for
fpointer_type
function
Header
if
ifelse
struct
switch
void
while

P. B {} F. 0.T.

a)   b)

The `helloBlinky_c1v1` folder we developed in *Chapter 1, A Practical Introduction to ARM®
CORTEX®* was quite small and could easily be described by a flowchart (try to sketch it), but as
programs become larger, their flowcharts become large and unwieldy. Handling complexity is
a common problem in all engineering disciplines and one that is solved by a technique called
hierarchical decomposition. This is a long name for something quite simple. It just means we
keep on subdividing complex designs into smaller and smaller parts until they become simple
enough to handle.

Writing a function

Functions (sometimes called subroutines) are used to hide the complexity of underlying
program statements, thereby presenting a more abstract view of the program. Abstraction is
commonplace in engineering; for example, we can think of a car as comprising subassemblies
that include body, engine, gearbox, suspension, and so on. The complexity within these
subassemblies is only important to those specialists such as designers, test engineers, and
technicians who need to interact with them. For example, the designers of the gearbox don't
need to concern themselves with the intricacies of the engine, they just need to know a few
important parameters. Functions provide a similar abstraction mechanism. We already met
the functions `LED_Initialize()`; `LED_On()`, and `LED_Off()` used to initialize and switch
the LEDs. We don't need to know exactly how these functions do their job but only how to
use them. C provides functions as a mechanism of achieving hierarchical decomposition. For
example, our `main()` function of `helloBinky_c2v1` is becoming a bit cluttered and difficult
to follow. To simplify the structure, the two for loops that simply introduce a delay could be
repackaged as a function called `delay()` that accepts one input arg (that determines the
length of the delay) and returns no output args (that is, `void`).

How to do it

1. Clone the `helloBlinky_c2v1` project to give `helloBlinky_c2v2`.

2. Edit `Blinky.c`, and define the function `delay()` by adding the following:

```
void delay (unsigned integer d) {
  unsigned integer i;

  for (i=0; i < d; i++)
    /* empty statement */ ;
}
```

3. It doesn't matter if the definition is placed before or after `main ( )`, but it shouldn't be nested inside `main()` (Note: functions defined inside other functions are called nested functions). Declare the function by including a function prototype declaration at the start of the program (that is, before the function is defined).

```
void delay (unsigned int);
```

4. Replace the statements:

```
for (i = 0; i < 1000000; i++)
    /* empty statement */ ;                /* Wait */
```

5. Call the following function:

```
delay (num_ticks);
```

6. Declare a new variable in `main()` and initialize it.

```
const unsigned int num_ticks = 500000;
```

7. The relevant changes are shown as follows (omitting boilerplate code):

```
void delay (unsigned int);          /* Func Prototype */

int main (void) {
  const unsigned int max_LEDs = 8;
  const unsigned int num_ticks = 500000;
  unsigned int num = 0;

  HAL_Init ( );      /* Init Hardware Abstraction Layer */
  SystemClock_Config ( );          /* Config Clocks */
  LED_Initialize ( );                  /* LED Init */

  for (;;) {                       /* Loop forever */
    LED_On (num);                  /* Turn LEDs on */
    delay (num_ticks);
    LED_Off (num);                 /* Turn LEDs off */
```

```
        delay (num_ticks);
        num = (num+1)%max_LEDs;    /* increment num (mod-8) */
    } /* end for */
} /* end main ( ) */

void delay (unsigned int d){        /* Function Def */
    unsigned int i;

    for (i = 0; i < d; i++)
    /* empty statement */ ;                    /* Wait */
    } /* end delay ( ) */
```

How it works...

Essentially, we've moved the `for` loop which implements the delay to within the function. The `for` loop itself is very similar to that used by `helloBlinky_c2v1`, except that the compare instruction used to terminate the loop now references the input argument d rather than a literal value (that is, 1000000).

```
for (i=0; i < d; i++) {

    ;

}
```

This is advantageous because it parameterizes the delay function, thereby allowing it to be used to implement different length delays, determined by the value of input argument d. An important feature of all programming languages is the mechanism they use to pass arguments to a function when it is called. There are two general models, called *pass-by-value* and *pass-by-reference*. The delay function call we've used here:

```
delay (num_ticks);
```

adopts a *pass-by-value* model. In this case, a copy of the variable `num_ticks` is passed to the `delay` function, and this copy can be referenced through the variable d. The statements inside the function can only access the variables declared within the function (that is, local to the function) and the input arguments. The function may change the value of the copy, but when the function terminates the copy (and the so-called automatic variables declared inside the function cease to exist). This model works fine in this case, because the function doesn't need to change the value of the variable `num_ticks` declared in `main()` (that is, the calling function).

All identifiers in C need to be declared before they are used. This is true of functions as well as variables (you may be catching onto the idea that C compilers don't tolerate surprises!), so functions should be declared before they are defined or called. A function declaration (also called a function prototype) includes the type of variable returned by the function, and the types of all the input args. C compilers accept the function definition as an implicit declaration and lazy programmers sometimes take advantage of this and omit the function prototype. But in this case, it must occur before the function is called. Nevertheless, it is considered good practice to include prototypes for all functions used. Function prototypes are usually placed at the beginning of the program or in a separate #include file. The prototype for our delay function looks like this:

```
void delay (unsigned integer);
```

[🔆 White space characters are ignored by the compiler; we only include them to make our code more readable.]

There's more...

If the delay function did need access to main functions variable, `num_ticks`, then it would need to access the memory location where `num_ticks` was stored. In this case, rather than passing a copy, we need to pass a reference (or so-called pointer) to the variable. C includes two special operators (* and &) for handling memory references. The ability to manipulate pointers as well as variables makes C a very powerful language, and it is a feature that is particularly useful for embedded systems programming. Consider the declaration:

```
unsigned int *ptr;
```

Here, `ptr` is the name of our variable, but in this case, it is preceded by the dereferencing operator * which tells the compiler it's a pointer variable, and so, the compiler must reserve enough memory to store an address. It also says the address will reference (that is, point to) an unsigned integer. When the pointer is declared and hasn't been assigned, we say the pointer is NULL (that is, its value cannot be guaranteed). To assign the pointer, we need to find the address of the variable `num_ticks`; the & operator achieves this. For example:

```
ptr = &num_ticks;
```

Let's consider another version of the delay function that doesn't declare the local variable `i`, but instead, employs a while loop that decrements the variable `num_ticks` declared in `main`. To do this, the function call to delay (within main) will need to pass a reference (or pointer) to `num_ticks`, and the `delay()` function will need to be told to expect a pointer to an unsigned integer as an input arg. Therefore, the function prototype will need to be changed to

```
void delay (unsigned int *);
```

and the function declaration itself becomes:

```
void delay (unsigned int *ptr) {

    while (*ptr > 0 )
    *ptr = (*ptr)-1;         /* Wait */
}
```

The `delay` function uses the dereferencing operator * whenever it needs to access the value pointed to by `ptr`. The following recipe (`helloBlinky_c2v3`) represents a version of `helloBlinky` that uses pointers:

```
void delay (unsigned int *);       /* Func Prototype */

int main (void) {
  const unsigned int max_LEDs = 8;
  const unsigned int wait_period = 500000;
  unsigned int *ptr;
  unsigned int num_ticks;
  unsigned int num = 0;

  HAL_Init ( );   /* Init Hardware Abstraction Layer */
  SystemClock_Config ( );             /* Config Clocks */
  LED_Initialize();                    /* LED Init */

  for (;;) {                         /* Loop forever */
    LED_On (num);                       /* LED on */
    num_ticks = wait_period;          /* (re)set delay */
    ptr = &num_ticks;                 /* assign pointer */
    delay (ptr);                  /* call delay function */
    LED_Off (num);                      /* LED off */
    num_ticks = wait_period;          /* (re)set delay */
    delay (ptr);                  /* call delay function */
    num = (num+1)%max_LEDs;       /* increment num (mod-8) */
  } /* end for */
} /* end main ( ) */

void delay (unsigned int *p){          /* Function Def */

  while (*p > 0 )
    *p = *p-1;                             /* Wait */

} /* end delay ( ) */
```

The preceding version of `helloBlinky` is just a vehicle for illustrating pointers, and the earlier recipe is preferable and easier to understand. So why are pointers used? Well, if our `delay` function needed access to many values, making the copies needed for *pass-by-value* would be time-consuming and impractical. This is particularly true when we come to consider passing arrays of data, strings (arrays of characters), and so on.

Writing to the console window

While a variant of the `helloBlinky` recipe is usually the first program introduced in most embedded tutorials, the first program found most C textbooks usually outputs the string "Hello World" to the screen. To run such a program on our evaluation board, we'll need to install a terminal emulation program on our PC host. **PuTTY®** `http://www.chiark.greenend.org.uk/~sgtatham/putty/`, an open source terminal emulation program is a good choice. We also need to connect the evaluation board to the PC's (COM) serial port. Most PCs and laptops are no longer fitted with 9-pin D-type (COM) ports, so you may need to purchase a USB to Serial Adaptor cable.

Getting ready

Follow these steps to install PuTTY, and connect the evaluation board to the PC's COM port:

1. If you're using a USB Serial Adaptor, then plug it into the laptop, and wait for the driver to be installed.

2. Open the Control Panel, and make a note of the COM port that has been allocated (you will need this later to configure PuTTY).

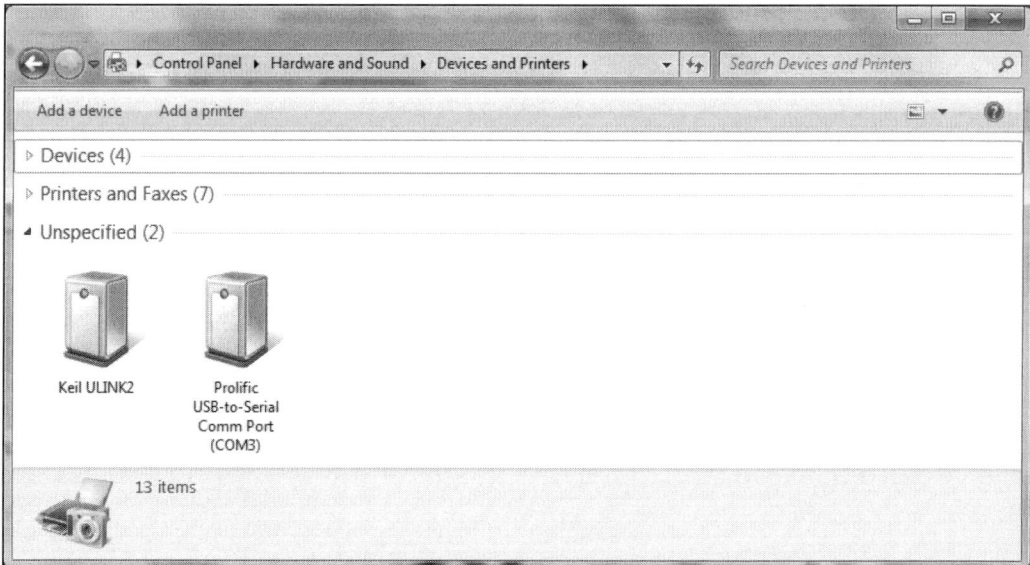

3. Connect the 9-Pin D-type UART1/3/4 connector on the evaluation board to the PC USB port, and ensure that the jumpers *J13* and *J14* are set to short pins 1 and 2 thereby selecting USART4. Pin 1 can be easily be identified by its square solder pad, easily visible on the underside of the board. Install PuTTY, and configure the serial connection to use the COM port you previously identified in Control Panel, configured to 115200 Baud, 8 data bits, 1 stop-bit, no parity or flow control.

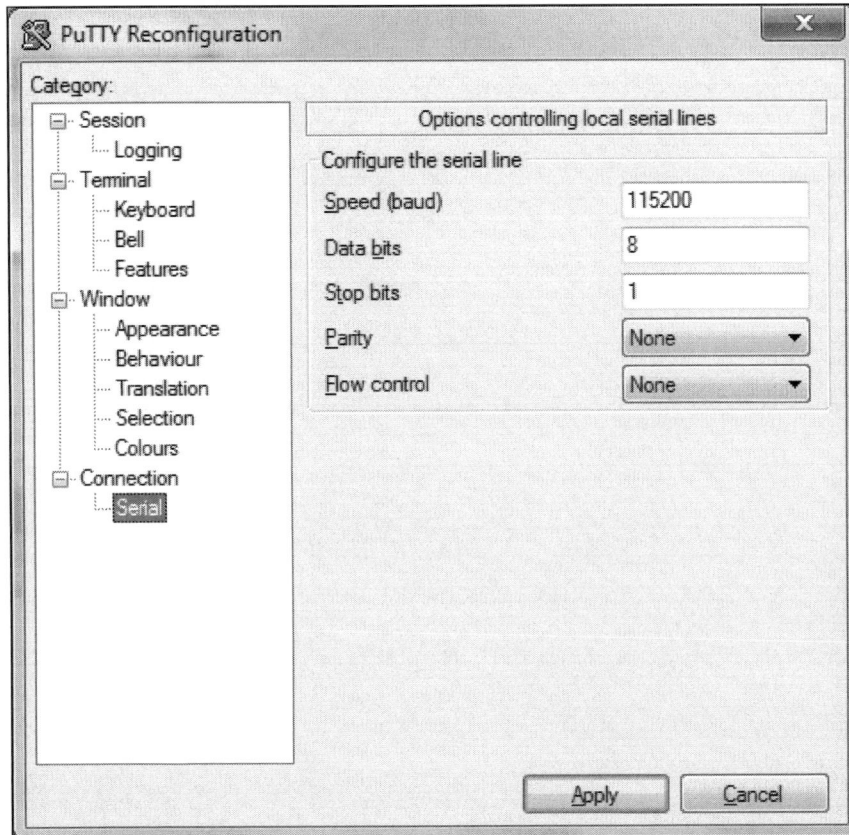

How to do it...

1. Create a new folder named `helloWorld`; invoke uVision5, and create a new project. Using the RTE manager, select the MCBSTM32F400 board, but don't check any of the board support tick boxes. Check **CMSIS → CORE**, **RTOS (API) → KeilRTX**, **Device → Startup**, and **Device → STM32Cube Framework (API) → Classic**. Click **Resolve** to automatically load any additional software components needed. Then exit by clicking on **OK**.

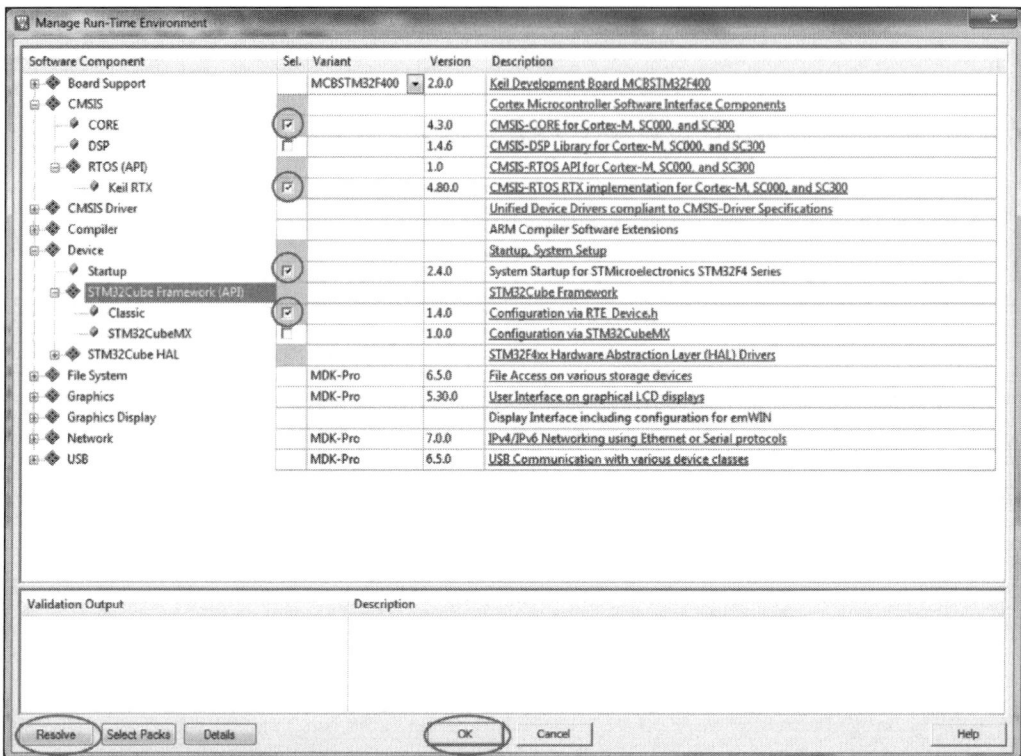

2. The source code for this project is divided between three source code files. Create a new file (**File →New...**), and enter the source code shown. Save the file (**File →SaveAs**) as `helloWorld.c`. The source file named `helloWorld.c` contains the main function in the project, illustrated using the folding editor feature to hide the boilerplate.

```
/*************************************************
 * Recipe:    helloWorld_c2v0
 * File:      helloWorld.c
 * Purpose:   Serial I/O Example
 *************************************************
 *
```

```
 * Modification History
 * 2014 Created
 * 03.12.15 Updated for uVision_5.17 & DFP_2.6.0
 *
 * Dr Mark Fisher, CMP, UEA, Norwich, UK.
 ***************************************************/

#include "stm32F4xx_hal.h"
#include "cmsis_os.h"
#include <stdio.h>
#include "Serial.h"

/* Function prototypes */
void wait(unsigned long delay);
extern void init_serial(void);
extern int sendchar(int c);
extern int getkey(void);

#ifdef __RTX
/* Function prototypes */
void wait(unsigned long delay);
extern void init_serial(void);
extern int sendchar(int c);
extern int getkey(void);

#ifdef __RTX

/*-------------------------------------------------
   System Clock Configuration
 *-------------------------------------------------*/

void SystemClock_Config(void) {

/*-------------------------------------------------
 *      wait
 *-------------------------------------------------*/
void wait (unsigned long delay){
unsigned long i;

   for (i = 0; i < delay; i++)
     ;
}

int main (void) {
  HAL_Init ();  /* Init Hardware Abstraction Layer */
  SystemClock_Config ();          /* Config Clocks */
  SER_Init ();
```

```
for (;;) {                            /* Loop forever */
  wait(1000000);
  printf("Hello World!\n");
}
}
```

3. In the project window, right-click on the `Source Group 1` folder, and add the source file `helloWorld.c` to the project.

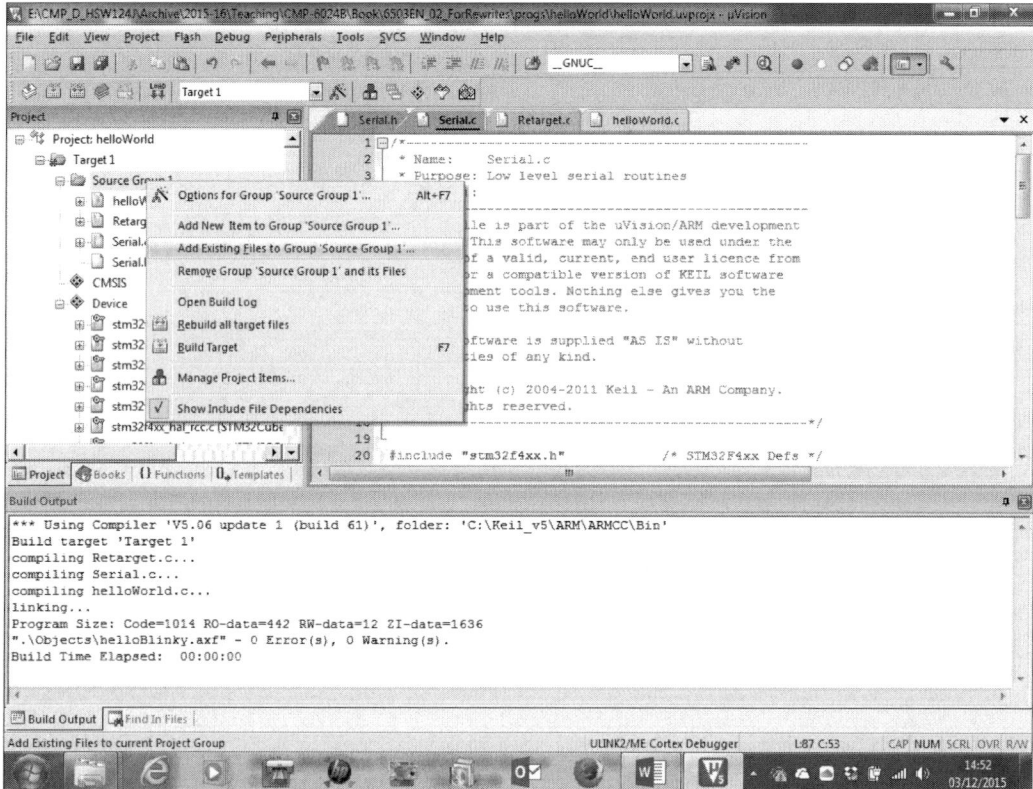

4. Create a new file, enter the following code, name the new file `Retarget.c`, and add it in the project. This source file redefines some functions used by C's standard input output library, `<stdio.h>`.

```
/*------------------------------------------------
 * Name:    Retarget.c
 * Purpose: 'Retarget' layer for target-
 *          dependent low level functions
 * Note(s):
 *------------------------------------------------
 * This file is part of the uVision/ARM
 * development tools.
```

```
 *-----------------------------------------------------*/

#include <stdio.h>
#include <rt_misc.h>
#include "Serial.h"

#pragma import(__use_no_semihosting_swi)

struct __FILE {
  int handle;
  /* Add whatever you need here */
};
FILE __stdout;
FILE __stdin;

int fputc(int c, FILE *f) {
  return (SER_PutChar(c));
}

int fgetc(FILE *f) {
  return (SER_GetChar());
}

int ferror(FILE *f) {
  /* Your implementation of ferror */
  return EOF;
}

void _ttywrch(int c) {
  SER_PutChar(c);
}

void _sys_exit(int return_code) {
label: goto label;  /* endless loop */
}
```

5. Create a new file, enter the SER_Init() function, name the new file Serial.c, and add it in the project.

```
/*-----------------------------------------------
 * Name:    Serial.c
 * Purpose: Low level serial routines
 * Note(s):
 *-----------------------------------------------
 * This file is part of the uVision/ARM
```

```
 * development tools.
 *-------------------------------------------------*/
#include "stm32f4xx.h"          /* STM32F4xx Defs */
#include "Serial.h"

#ifdef __DBG_ITM
volatile int32_t ITM_RxBuffer;
#endif
/*-----------------------------------------------
 * SER_Init:  Initialize Serial Interface
 *-------------------------------------------------*/
void SER_Init (void) {
#ifdef __DBG_ITM
  ITM_RxBuffer = ITM_RXBUFFER_EMPTY;

#else
  RCC->APB1ENR  |= (1UL << 19); /* Enable
                                      USART4 clock */
  RCC->APB2ENR  |= (1UL <<  0); /* Enable
                                       AFIO clock */
  RCC->AHB1ENR  |= (1UL <<  2); /* Enable
                                      GPIOC clock */
  GPIOC->MODER  &= 0xFF0FFFFF;
  GPIOC->MODER  |= 0x00A00000;
  GPIOC->AFR[1] |= 0x00008800;  /* PC10 UART4_Tx,
                             PC11 UART4_Rx (AF8) */

  /* Configure UART4: 115200 baud @ 42MHz, 8 bits,
                          1 stop bit, no parity */
  UART4->BRR = (22 << 4) | 12;
  UART4->CR2 = 0x0000;
  UART4->CR1 = 0x200C;
#endif
}
```

6. Add the functions SER_getc() and SER_putc() to Serial.c

```
/*-----------------------------------------------
 * SER_PutChar:  Write a char to Serial Port
 *-------------------------------------------------*/
int32_t SER_PutChar (int32_t ch) {
#ifdef __DBG_ITM
  int i;
  ITM_SendChar (ch & 0xFF);
  for (i = 10000; i; i--)
```

```
      ;
#else
  while (!(UART4->SR & 0x0080));
    UART4->DR = (ch & 0xFF);
#endif

  return (ch);
}

/*-------------------------------------------------
 * SER_GetChar:  Read a char from Serial Port
 *-----------------------------------------------*/
int32_t SER_GetChar (void) {
#ifdef __DBG_ITM
  if (ITM_CheckChar())
    return ITM_ReceiveChar();
#else
  if (UART4->SR & 0x0020)
    return (UART4->DR);
#endif
  return (-1);
}
```

7. Create a new file, enter the following code, name the file `Serial.h`, and add it to the project. This is the header file that declares the function prototypes for `Serial.c`

```
/*-------------------------------------------------
 * Name:    Serial.h
 * Purpose: Low level serial definitions
 * Note(s):
 *-----------------------------------------------*/
#ifndef __SERIAL_H
#define __SERIAL_H

extern void SER_Init     (void);
extern int  SER_GetChar  (void);
extern int  SER_PutChar  (int c);

#endif
```

8. Configure PuTTY as shown in part a) of the following image. Build, download, and run the program to achieve the output shown in b)

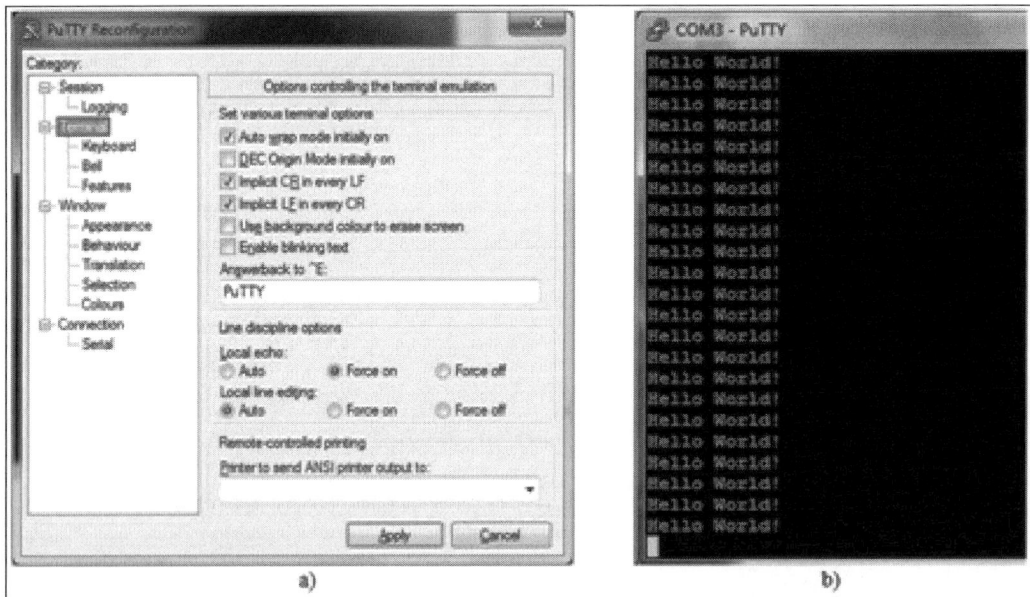

a)   b)

How it works...

The evaluation board and PC communicate by exchanging data using an RS232 serial Input/Output (I/O) connection (http://en.wikipedia.org/wiki/RS-232). RS232 is a 2-wire full-duplex communications standard. PuTTY manages the protocol at the PC, but we are responsible for the evaluation board. To use serial I/O, we need to configure the microcontroller's **Universal Synchronous/Asynchronous Receiver/Transmitter** (**USART**). We can do this by including a peripheral driver applications interface (API) in our project. uVision5's RTE manager includes a suitable API, but this provides many more features than we need for our simple *helloWorld* recipe. So, for the time being, we'll use the simpler driver named Serial.c shown in step 4 and step 5 that ARM shipped with uVision4. File Serial.c comprises three functions SER_Init(), SER_PutChar(), and SER_GetChar(). The function SER_Init() is the first function called by main(). It initializes the USART peripheral by writing values to its registers so that it is configured to mirror the channel setup in PuTTY (that is, 115200 baud, 8 data-bits, 1 stop-bit). These parameters are critical. The baud rate is derived from the Peripheral Clock, and in turn the System Clock, so any change in the clock configuration will affect the baud rate. The baud rate is set by the value we write to the **Baud Rate Register** (**BRR**). Reference manual *RM0090* (www.st.com) describes this as calculated by

$$\text{Tx/Rx baud} = \frac{f_{clk}}{8(2 \times \text{OVERS}) \times \text{USARTDIV}}$$

Rearranging the preceding formula, with *OVER8 = 1* (since we're using 8 x oversampling) and *fclk = 42 MHz* we get:

$$\text{USARTDIV} = \frac{42 \times 10^6}{16 \times 115200} = 22.78 = 22\frac{12}{16}$$

The other two functions read and write characters from/to the USART (these perform the low-level I/O). We'll discuss this in more detail in *Chapter 3, Assembly Language Programming.*

Any program that wishes to use the services that `Serial.c` provides must include its function prototype. To facilitate this, the prototypes are declared in a so-called header file called `Serial.h` shown in step 6, and included in the program using a `#include` preprocessor directive (for example, see line 15 of `main.c`). If we look closely at `Serial.h`, we see the prototypes are preceded by the qualifier extern. This is a message to the compiler that the functions are defined in another file (that is, not `main.c`), and the function call reference must be resolved later by the linker. We can also see that the prototype declarations are enclosed within a conditional preprocessor statement, that is:

```
#ifndef __SERIAL_H
#define __SERIAL_H

/* function prototypes */

#endif
```

This ensures that the code enclosed within the conditional preprocessor statement is included in the project only once, even though both, `main.c` and `Serial.c`, include the statement:

```
#include "serial.h"
```

The `main()` function calls `printf()` to output the string `"Hello World\n"`. The string `"Hello World\n"` is stored as a sequence of characters terminated by a NULL character. C interprets `'\n'` as a newline character, but the actual ASCII code (http://en.wikipedia. org/wiki/ASCII) used to represent newline varies between operating systems; so to cover all eventualities, we can configure PuTTY as shown in step 7.

The function `printf()` is defined in C's standard input output library `<stdio.h>`. This function calls `fputc()`, which is also defined in `<stdio.h>`, but redefined in `Retarget.c`. So it calls `SER_PutChar()` to send the characters to the USART. Most microcontrollers use this technique to allow them to make use of the C library functions `printf()` and, as we'll see later, `scanf()` too.

File `Retarget.c` also uses the preprocessor directive `#pragma`, which is used to specify machine- or operating system-specific compiler features. In this case, the directive is used to disable semihosting. Semihosting is a mechanism that allows ARM targets to communicate with a host computer using the JTAG interface. Semihosting can be used with the function `trace_printf()`, to enable debug statements to write to the output window of the IDE. Obviously, we can achieve similar functionality using the COM port and PuTTY.

Writing to the GLCD

Although the LED flashing programs we've written so far have served to provide a tutorial introduction to C, you are probably ready for something a little more exciting. The **Graphic LCD (GLCD)** touchscreen provides an interactive interface based on a 320 x 240 pixel color display. Keil provides a library of functions to write characters and bit-mapped graphics to the screen.

Getting ready

1. Create a new folder and rename it `helloLCD_c2v0`. Invoke uVision5, and create a new project.

2. After selecting the target device (STM32F407IGHx), use the RTE manager to select the MCBSTM32F400 target board, and check the following software components: **Board Support → Graphic LCD, CMSIS → CORE, CMSIS → RTOS (API) → KeilRTX, Device → Startup, Device → STM32Cube Framework (API) → Classic**. Finally, left-click on **Resolve** and **OK**.

How to do it...

1. Create a new C source file called `helloLCD.c`, and enter the following statements. Although hidden by a fold, don't forget to add the boilerplate code we described in the recipe `helloBlinky_c2v0`.

```
/*-----------------------------------------------
 * Recipe:  helloLCD_c2v0
 * Name:    helloLCD.c
 * Purpose: LCD Touchscreen Demo
 *-----------------------------------------------
 *
 * Modification History
 * 06.02.14 Created
 * 08.12.15 Updated (uVision5.17 & DFP2.6.0)
 *
 * Dr Mark Fisher, CMP, UEA, Norwich, UK
 *-----------------------------------------------*/
```

```
#include "stm32f4xx_hal.h"
#include "GLCD_Config.h"
#include "Board_GLCD.h"

#define wait_delay HAL_Delay

extern GLCD_FONT    GLCD_Font_6x8;
extern GLCD_FONT    GLCD_Font_16x24;

#ifdef __RTX
```

```
/* Function Prototypes */
void SystemClock_Config(void);

/**
 * System Clock Configuration
 */
void SystemClock_Config(void) {
```

```
/**
 * Main function
 */
int main ( ) {
  unsigned int count;

  HAL_Init ();    /* Init Hardware Abstraction Layer */
  SystemClock_Config ();           /* Config Clocks */

  GLCD_Initialize();
  GLCD_SetBackgroundColor (GLCD_COLOR_WHITE);
  GLCD_ClearScreen ();
  GLCD_SetBackgroundColor (GLCD_COLOR_BLUE);
  GLCD_SetForegroundColor (GLCD_COLOR_WHITE);
  GLCD_SetFont (&GLCD_Font_16x24);
  GLCD_DrawString (0, 0*24, " CORTEX-M4 COOKBOOK ");
  GLCD_DrawString (0, 1*24, "  PACKT Publishing  ");
  GLCD_SetBackgroundColor (GLCD_COLOR_WHITE);
  GLCD_SetForegroundColor (GLCD_COLOR_BLUE);

  for (;;) {
    if (count==0)
      GLCD_DrawString (0, 2*24, "     Hello LCD!     ");
```

```
          else
            GLCD_DrawString (0, 2*24, "                              ");

          wait_delay( 100 );
          count = ( count+1 )%2;
       }  /* end for */
    }
```

2. Build, download, and run the program.

How it works...

The functions beginning `GLCD_` are defined in the file `GLCD_MCBSTM32F400.c`. We need to open this, and read the comments in the function headers to understand how to use them. The header file `Boadd_LCD.h` that is included by the pre-processor contains the function prototype declarations. The header file `GLCD_Config.h` provides macros that define named colors (like, `GLCD_COLOR_BLACK`) and constants such as `GLCD_WIDTH / HEIGHT`. `GLCD_MCBSTM32F400.c` is the latest in a series of GLCD drivers provided by Keil, and it represents a CMSIS v2.0-compliant revision of earlier versions.

The function `GLCD_DrawString (uint32_t x, uint32_t y, const char *str)` declared in file `Board_GLCD.h` takes three input arguments (args). The first two position the text on the screen, and the last arg is a pointer to an array of characters to be written (usually a literal value defined using quotes " " in the function call). Before calling `GLCD_DrawString ( )`, we must first set the character font to be used by the calling function, `GLCD_SetFont (GLCD_FONT *font)`, and pass a pointer to the font used. There are two font sizes defined in file `GLCD_Fonts.c`. An array of characters terminated by a NULL character is called a string. You may wonder why we didn't need to use the & operator to recover an address and assign a pointer as we illustrated earlier. The short answer is that arrays are always referenced using pointers, so there is no need, but we'll discuss the matter further in *Chapter 3, Assembly Language Programming*.

The macro definition `#define wait_delay HAL_Delay` provides a pseudonym for the function `HAL_Delay ( )` declared in the file `st32f4xx_HAL.h`. This is a more accurate delay based on a timer rather than an instruction loop.

Creating a game application – Stage 1

Now that we can write characters to the GLCD screen, some interesting possibilities present themselves. The first one to consider is a simple character-based game application known as PONG. Pong was one of the first arcade video games featuring 2D graphics, originally marketed by ATARI Inc. (`http://en.wikipedia.org/wiki/Pong`). We'll develop the game in stages, as this is a good development strategy. We'll start by describing a simple recipe named Bounce with limited functionality. The idea of this recipe is just to animate a ball so that it appears to bounce around the screen. Provided we can redraw the ball more than 25 times a second (25 Hz), it will appear to move. The ball is represented by a character bitmap.

How to do it...

1. As usual, we'll start our development by making a new folder named `helloBounce_c2v0`. Create a project, and configure the RTE to include software support for the Graphic LCD board feature (that is, clone the folder `helloLCD_c2v0`, from the previous recipe).

2. Create a new file, enter the following code, name the file `helloBounce.c`, and include it in the project.

    ```
    /*-------------------------------------------------
     * Recipe:   helloBounce_c2v0
     * Name:     helloBounce.c
     * Purpose:  Pong Game Prototype
     *-------------------------------------------------
     *
     * Modification History
     * 06.02.14 Created
     * 08.12.15 Updated uVision5.17 + DFP2.6.0
     *
     * Dr Mark Fisher, CMP, UEA, Norwich, UK
     *-------------------------------------------------*/

    #include "stm32f4xx_hal.h"
    #include "GLCD_Config.h"
    #include "Board_GLCD.h"

    #define wait_delay HAL_Delay

    /* Globals */
    extern GLCD_FONT    GLCD_Font_16x24;

    #ifdef __RTX
    ```

```
/* Function Prototypes */
void SystemClock_Config(void);

/**
  * System Clock Configuration
  */
void SystemClock_Config(void) {
```
_____

```
/**
  * Main function
  */
int main (void) {
  unsigned int dirn = 1;
  /* initial ball position */
  unsigned int x = (GLCD_WIDTH-GLCD_Font_16x24.width)/2;
  unsigned int y = (GLCD_HEIGHT-
                                GLCD_Font_16x24.height)
  unsigned long num_ticks = 5;

  HAL_Init ( );
  SystemClock_Config ( );

  GLCD_Initialize();
  GLCD_SetBackgroundColor (GLCD_COLOR_WHITE);
  GLCD_ClearScreen ();
  GLCD_SetForegroundColor (GLCD_COLOR_BLUE);
  GLCD_SetFont (&GLCD_Font_16x24);
  GLCD_DrawChar (x, y, 0x81);               /* Draw Ball */

  for (;;) {                               /* superloop */
    wait_delay(num_ticks);         /* update ball pstn */
    /* add code to update ball position
       and check for collisions here */
    GLCD_DrawChar (x, y, 0x81);       /* Redraw Ball */
  } /* end for */
}
```

3. Build the project (just to check that there are no syntax errors).

   Include the following code fragment in the superloop of bounce.c. This code updates the position of the ball on each iteration.

```
/* update ball position */
switch (dirn) {
  case 0: x++;
          break;
```

```
     case 1: x++;
             y--;
             break;
     case 2: y--;
             break;
     case 3: x--;
             y--;
             break;
     case 4: x--;
             break;
     case 5: x--;
             y++;
             break;
     case 6: y++;
             break;
     case 7: x++;
             y++;
  }
```

Extend the superloop of bounce.c by including the code fragment that is designed to detect collisions between the ball and the edges of the screen. The ball direction is changed accordingly when a collision occurs.

```
   /* check collision with vertical screen edge */
   if ((x==0) ||
       (x==GLCD_WIDTH-GLCD_Font_16x24.width) ) {
     switch (dirn)
     {
       case 0: dirn = (dirn+4)%8;
               break;
       case 1: dirn = (dirn+2)%8;
               break;
       case 3: dirn = (dirn+6)%8;
               break;
       case 4: dirn = (dirn+4)%8;
               break;
       case 5: dirn = (dirn+2)%8;
               break;
       case 7: dirn = (dirn+6)%8;
               break;
     }
   }
   /* check collision with horizontal screen edge */
     if ((y==0) ||
         (y==GLCD_HEIGHT-GLCD_Font_16x24.height) ) {
```

```
switch (dirn) {
  case 1: dirn = (dirn+6)%8;
          break;
  case 2: dirn = (dirn+4)%8;
          break;
  case 3: dirn = (dirn+2)%8;
          break;
  case 5: dirn = (dirn+6)%8;
          break;
  case 6: dirn = (dirn+4)%8;
          break;
  case 7: dirn = (dirn+2)%8;
          break;
  }
}
```

4. Build the project; download and run the program. Observe the ball bouncing around the screen. Note that the argument passed to the function delay( ) controls the ball's speed. Experiment by changing the value.

How it works...

The direction of the ball is encoded by a number, 0-7, as shown in the following diagram. The ball's behavior when it strikes the edge of the screen depends on the angle of collision (in a similar manner to those on a pool table). Adding a value to the direction code (modulo-8) will change the ball's direction.

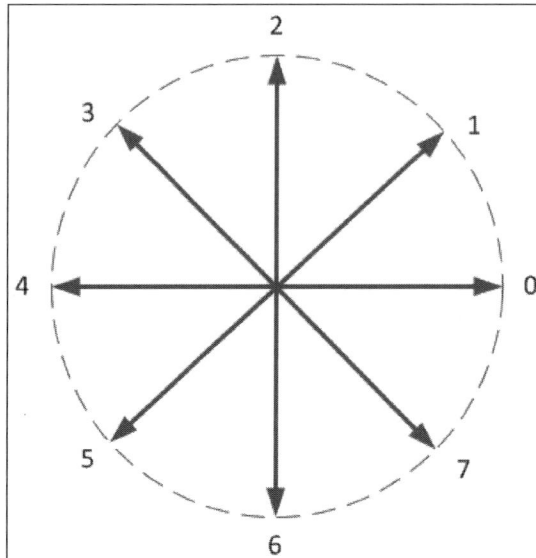

Characters we write to the GLCD are represented by bitmaps. Each character bitmap is represented as a *16 x 24* grid of cells. Each row of cells in the grid is encoded as two bytes, represented in hexadecimal. For example, the bitmap representation of the '&' character is illustrated in the following image:

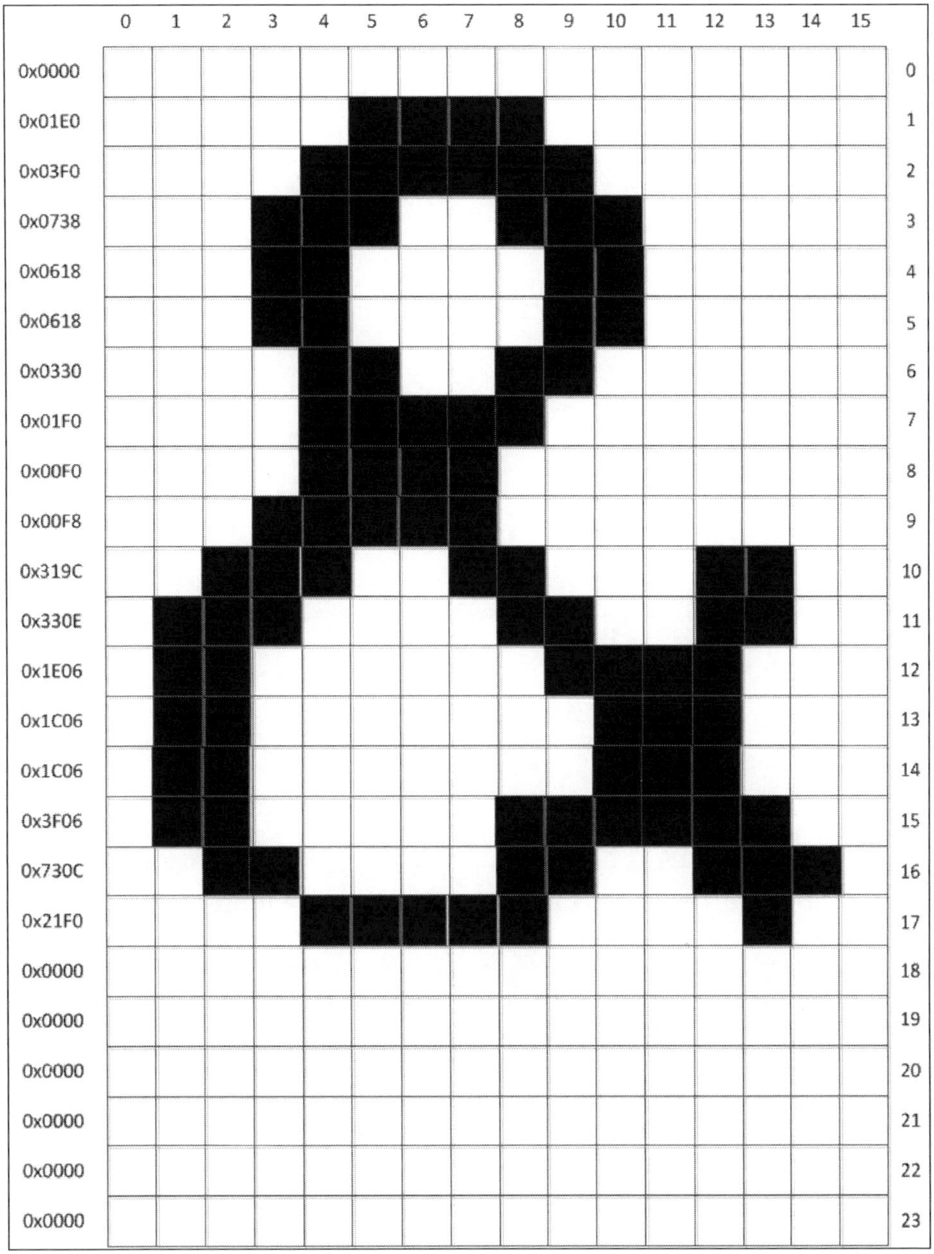

	0 1 2 3 4 5 6 7 8 9 10 11 12 13 14 15	
0x0000		0
0x01E0		1
0x03F0		2
0x0738		3
0x0618		4
0x0618		5
0x0330		6
0x01F0		7
0x00F0		8
0x00F8		9
0x319C		10
0x330E		11
0x1E06		12
0x1C06		13
0x1C06		14
0x3F06		15
0x730C		16
0x21F0		17
0x0000		18
0x0000		19
0x0000		20
0x0000		21
0x0000		22
0x0000		23

A good bitmap representation for the ball is a 'Circle – Full' character ($0x81 = 129_{10}$). We can display this character in any position on the GLCD screen using the function `GLCD_DrawChar()`. This function takes three args: screen coordinates (x, y), and the ASCII code for the character. The code fragment

```
GLCD_SetFont (&GLCD_Font_16x24);
GLCD_DrawChar (0, 0, 0x81);
```

will draw the ball in the top-left corner of the screen. GLCD_DrawChar ( ) interprets the ASCII character code as an index into GLCD_Font_16x24. The 'Circle – Full' character is the 97th character (of a total of 112) stored in the array named GLCD_ `Font_24x16`. Parameters for the font are stored in the file `GLCD_Fonts.c`.

```
GLCD_FONT GLCD_Font_16x24 = {
    16,                          ///< Character width
    24,                          ///< Character height
    32,                          ///< Character offset
    112,                         ///< Character count
    Font_16x24_h                 ///< Characters bitmaps
};
```

If we add the Character offset (32_{10}) to the character's position in the character bitmap (97_{10}), we get its code (129_{10}).

Finally, since the character bitmap is not declared in `bounce.c`, we need to tell the compiler what type `Font_16x24_h` is, and that it is declared elsewhere. The statement

```
extern GLCD_FONT        GLCD_Font_16x24;
```

in file `bounce.h` achieves this. This file also uses the #define preprocessor directive to declare global constants (such as `CHAR_W` and `CHAR_H`). Conventionally, these are capitalized.

The superloop comprises statements that animate the ball by updating its position (x,y) and redrawing the bitmap. Position updates depend on direction (encoded as, 0,1,2,3,4,5,6, or 7). These eight cases are identified by the switch statement in step 7 of the recipe. Our trusty delay function provides some control over the speed of the ball. Further code in the superloop checks for collisions between the ball and the vertical and horizontal edges of the screen, and updates the balls direction appropriately. The last statement in the superloop is a further call to the function `GLCD_Draw_Char()` to redraw the ball in its new location. Because the bitmap represents a solid circle shape surrounded by a border of background pixels, and since the ball position is only incremented by a single pixel each time there, is no need to erase the ball before it is redrawn.

Creating a game application – Stage 2

This prototype extends the one described in the previous section to make a single player game that includes a '*paddle*' drawn on left-hand edge of the screen. The position of the paddle is determined by a potentiometer (ADC1) fitted to the evaluation board that provides a voltage input to the Analog-Digital (A-D) Converter.

1. Begin by creating a new folder named `helloPong_c2v0`, and within this, a new project. Configure the RTE to include board support software components for the Graphic LCD (API) and A/D Converter (API). Alternatively, clone the folder `helloBounce_c2v0`, from the previous recipe and modify the RTE. Use Resolve to automatically load any missing libraries.

2. Copy `helloBounce.c` and `helloBounce.h` from the previous recipe, rename them `helloPong.c` and `helloPong.h`, and include these in your project. Change the `#include` in `helloPong.c`, and replace `helloBounce.h` with `helloPong.h`. Build the program and test it as before.

3. Add `#include "Board_ADC.h"` and call `ADC_Initialize()` in `main()`.

4. Add a function named `update_ball()`, and move the code concerned with updating the ball's position and collision detection into the body of the function. This tidies up the superloop and makes the main function much easier to read.

5. Define constants and declare global data structures in `helloPong.c` to hold the position of the ball, paddle, and information about the Game.

```
#define wait_delay HAL_Delay
#define WIDTH    GLCD_WIDTH
#define HEIGHT   GLCD_HEIGHT
#define CHAR_H   GLCD_Font_16x24.height
/* Character Height (in pixels) */
#define CHAR_W   GLCD_Font_16x24.width
/* Character Width (in pixels)  */
#define BAR_W    6              /* Bar Width (in pixels) */
#define BAR_H    24             /* Bar Height (in pixels) */
#define T_LONG   1000                  /* Long delay */
#define T_SHORT 5                    /* Short delay */

typedef struct {
      int dirn;
      int x;
      int y;
} BallInfo;

typedef struct {
    int x;
    int y;
```

```
   } PaddleInfo;

   typedef struct {
     unsigned int num_ticks;
     BallInfo ball;
     PaddleInfo p1;
   } GameInfo;

   /* Function Prototypes */
   void game_Initialize(void);
   void update_ball (void);
   void update_player (void);
   void check_collision (void);
```

6. Declare a global variable in file `pong.c`:

    ```
    GameInfo thisGame;
    ```

    [💡 The ball's position is now accessed as thisGame.ball.x.]

7. Declare the function `game_Initialize()`. This function initializes the values of the global variables.

    ```
    /*--------------------------------------------------
     *   game_Init()
     *   Initialize some game parameters.
     *--------------------------------------------------*/
    void game_Initialize(void)
       init_pstn.dirn = 1;
       init_pstn.x = WIDTH-CHAR_W)/2;
       init_pstn.y = (HEIGHT-CHAR_H)/2;
       thisGame.ball = init_pstn;
       thisGame.p1.x = 0;
       thisGame.p1.y = 0;
       thisGame.num_ticks = T_SHORT;
    }
    ```

8. Create a new function named `check_collision()`, and copy the code concerned with collision detection into this function. Modify the function `check_collision()` to check for collisions between the ball and the paddle as well as collisions between the ball and screen edge.

    ```
    /*--------------------------------------------------
     *   check_collision(void)
     *   check for contact between ball and screen
    ```

```
*   edges/bat and change direction accordingly
*--------------------------------------------------*/
void check_collision(void) {
  /* check collision with RH vertical screen
                                  edge OR P1 paddle */
  if ((thisGame.ball.x == BAR_W) ||
      thisGame.ball.x == (WIDTH-CHAR_W)) {

    switch (thisGame.ball.dirn) {
      case 0: thisGame.ball.dirn =
                (thisGame.ball.dirn+4)%8;
              break;
      case 1: thisGame.ball.dirn =
                (thisGame.ball.dirn+2)%8;
              break;
      case 3: if ( (thisGame.ball.y >=
                    thisGame.p1.y-CHAR_H) &&
                   (thisGame.ball.y <=
                     (thisGame.p1.y+BAR_H)) )
                thisGame.ball.dirn =
                  (thisGame.ball.dirn+6)%8;
              else
                /* empty statement */
                break;
      case 4: if ( (thisGame.ball.y >=
                    thisGame.p1.y-CHAR_H) &&
                   (thisGame.ball.y <=
                     (thisGame.p1.y+BAR_H)) )
                thisGame.ball.dirn =
                  (thisGame.ball.dirn+4)%8;
              else
                /* empty statement */;
              break;
      case 5: if ( (thisGame.ball.y >=
                    thisGame.p1.y-CHAR_H) &&
                   (thisGame.ball.y <=
                     (thisGame.p1.y+BAR_H)) )
                thisGame.ball.dirn =
                  (thisGame.ball.dirn+2)%8;
              else
                /* empty statement */;
              break;
      case 7: thisGame.ball.dirn =
                (thisGame.ball.dirn+6)%8;
```

```
                    break;
          }
      }
      /* check collision with horizontal screen edge */
      if ((thisGame.ball.y < 0) ||
            thisGame.ball.y > (HEIGHT-CHAR_H)) {
        switch (thisGame.ball.dirn) {
          case 1: thisGame.ball.dirn =
                    (thisGame.ball.dirn+6)%8;
                  thisGame.ball.y++;
                  break;
          case 2: thisGame.ball.dirn =
                    (thisGame.ball.dirn+4)%8;
                  thisGame.ball.y++;
                  break;
          case 3: thisGame.ball.dirn =
                    (thisGame.ball.dirn+2)%8;
                  thisGame.ball.y++;
                  break;
          case 5: thisGame.ball.dirn =
                    (thisGame.ball.dirn+6)%8;
                  thisGame.ball.y--;
                  break;
          case 6: thisGame.ball.dirn =
                    (thisGame.ball.dirn+4)%8;
                  thisGame.ball.y--;
                  break;
          case 7: thisGame.ball.dirn =
                    (thisGame.ball.dirn+2)%8;
                  thisGame.ball.y--;
                  break;
        }
      }
  }
```

9. Add the following code fragment to the function `update_ball()`:

```
/* reset position */
if (thisGame.ball.x<BAR_W) {
  wait_delay (T_LONG);
  /* Erase Ball */
  GLCD_DrawChar( thisGame.ball.x, thisGame.ball.y, ' ');
  thisGame.ball = init_pstn;
}
```

10. Define `GLCD_customFont_16x24` in the file `GLCD_customFont.c`, and add this to the project.

```
#include "Board_GLCD.h"

static const uint8_t customFont_16x24_h[] = {
/* PONG PADDLE */
   0x00, 0x3F, 0x00, 0x3F, 0x00, 0x3F, 0x00, 0x3F,
   0x00, 0x3F, 0x00, 0x3F, 0x00, 0x3F, 0x00, 0x3F,
   0x00, 0x3F, 0x00, 0x3F, 0x00, 0x3F, 0x00, 0x3F,
   0x00, 0x3F, 0x00, 0x3F, 0x00, 0x3F, 0x00, 0x3F,
   0x00, 0x3F, 0x00, 0x3F, 0x00, 0x3F, 0x00, 0x3F,
   0x00, 0x3F, 0x00, 0x3F, 0x00, 0x3F, 0x00, 0x3F,
};

GLCD_FONT GLCD_customFont_16x24 = {
   16,                      ///< Character width
   24,                      ///< Character height
   0,                       ///< Character offset
   1,                       ///< Character count
customFont_16x24_h          ///< Characters bitmaps
};
```

11. Define the function `update_player()` by adding the following code fragment:

```
/*---------------------------------------------
* update_player(unsigned int *)
* Read the ADC and draw the player 1's paddle
*---------------------------------------------*/
void update_player(void) {

   int adcValue;
   static int lastValue = 0;

   ADC_StartConversion();
   adcValue = ADC_GetValue ();
   adcValue = (adcValue >> 4) * (HEIGHT-BAR_H)/256;
   /* Erase Paddle */
   GLCD_DrawChar (0, lastValue, ' ');
   /* Draw Paddle */
   GLCD_SetFont (&GLCD_customFont_16x24);
   GLCD_DrawChar (0, adcValue, 0x00 );
   GLCD_SetFont (&GLCD_Font_16x24);
   lastValue = adcValue;
   thisGame.p1.y = adcValue;
}
```

12. Build the project, download, and run.

There's more...

1. We can tidy the code by moving the function prototype and data structure declarations to a header file called `helloPong.h`, and include this in `pong.c` with a `#include` preprocessor directive.

```
/*-----------------------------------------------------
 * Recipe:  helloPong_c1v0
 * Name:    helloPong.h
 * Purpose: pong function prototypes and defs
 *-----------------------------------------------------
 *
 * Modification History
 * 06.02.14 Created
 * 09.12.15 Updated (uVision5.17 + DFP2.6.0)
 *
 * Dr Mark Fisher, CMP, UEA, Norwich, UK
 *-----------------------------------------------------*/
#ifndef _PONG_H
#define _PONG_H

#define wait_delay HAL_Delay
#define WIDTH GLCD_WIDTH
#define HEIGHT       GLCD_HEIGHT
#define CHAR_H  GLCD_Font_16x24.height
/* Character Height (in pixels) */
#define CHAR_W  GLCD_Font_16x24.width
/* Character Width (in pixels)  */
#define BAR_W    6          /* Bar Width (in pixels) */
#define BAR_H    24         /* Bar Height (in pixels) */
#define T_LONG  1000               /* Long delay */
#define T_SHORT 5                  /* Short delay */

typedef struct {
    int dirn;
    int x;
    int y;
  } BallInfo;

typedef struct {
    int x;
    int y;
} PaddleInfo;

typedef struct {
```

```
        unsigned int num_ticks;
        BallInfo ball;
        PaddleInfo p1;
    } GameInfo;

    /* Function Prototypes */
    void game_Initialize(void);
    void update_ball (void);
    void update_player (void);
    void check_collision (void);

    #endif /* _PONG_H */
```

2. The function declarations `game_Initialize()`, `update_ball()`, `update_player()`, and `check_collision()` can be moved to a file called `pong_utils.c`, which shares the header `pong.h`.

How it works...

The data structures defined within `pong.h` define three new *compound* data types which build on the primitive types such as char, integer, and so on, which are part of the language. A global variable `thisGame` stores all the data used in the application. The main file `helloPong.c` is shown in step 6. New functions `game_Initialize()`, `update_ball()`, `update_player()`, and `check_collision()` have been defined within the file `pong_utils.c` (and delay has also been moved) to declutter main and improve the readability of the code. The function prototypes are shown in step 9.

The function `game_Initialize( )` writes the initial values to the global structs, `gameInfo` and `init_pstn()`. The function `update_player()` (step 10) reads the A-D converter, and draws the paddle. Since the paddle may move in large increments, we must explicitly erase the paddle, and redraw it in a new position. The `static` qualifier is used to ensure that the variable `lastValue` persists after the function has terminated (that is, it behaves rather like a global variable, although its scope is local to the function). It is important to understand the scoping rules for variables. Variables declared within a function (so-called automatic variables) can only be changed by assignments within the function. But variables declared outside a function have global scope, and can be accessed by any function declared within the same file. The variable `gameInfo` is a global variable and can be accessed by any function declared in `helloPong.c`, and because of the extern declaration, by any function declared in `pong_utils.c`.

The functions named check_collision() and update_ball() are similar to those described in the previous section but with some important additions. When the ball moves in directions 3, 4, or 5, we need to check for a collision with the paddle; modifications necessary to achieve this are shown in step 8. If the ball fails to make contact with the paddle, then a clause in update_ball() holds the ball in its current position for a few seconds, and then restarts the game (see step 9).

The paddle itself can be drawn by declaring our own 'paddle' character bitmap in file GLCD_customFont.c, and by using GLCD_DrawChar() to render it to the screen. The code for checking collisions needs to be extended to include collisions between the ball and the inner vertical edge of the paddle. These can only occur when the ball direction is from right to left (that is, direction codes 3, 4 and 5). We'll need variables to represent the position of the paddle (as we do in case of the ball). As we now have quite a few variables, it's a good opportunity to introduce a data structure that can be used to group them together. The C struct provides us with a mechanism for achieving this. Information about the ball are declared in a struct called ballInfo. The information associated with the paddle is declared in paddleInfo and that about the game in gameInfo, within helloPong.h

Debugging your code using print statements

This section deals with debugging. Errors fall into two classes, compilation errors and run-time errors. Compilation errors arise when we compile our programs, and the compiler parses each of the statements to produce executable code. Syntactic errors such as a missing semi-colon or forgetting to declare a variable before assigning it will produce a compilation error. Luckily, uVision5 highlights and checks the syntax of our programs as we type. So, many problems that would have gone undetected in the past are now brought to our attention before compilation. When errors do occur, they are printed in the output window together with details of the file and the line number where the error occurred. In addition to errors, the compiler will also issue warnings relating to unusual conditions in the code that might be indicative of a problem. It's a good plan to treat warnings as errors, and track down their source. Further information about compiler diagnostic messages is in the Compiler User Guide that can be found in the Tool's Users Guide accessed by the Books tab of the IDE.

Runtime errors are generally harder to fix than those that occur during compilation. Adopting a good development strategy can minimize problems, or at least enable the problem to be quickly isolated. Larger programs are never written all at once, they always build on previously tested functions. The most straightforward way to debug a program is by inserting statements that print to the Graphic LCD screen, using GLCD_DisplayString( ).

How to do it...

To output the values of variables that are used by the program, we need to convert integer, unsigned integer, and such values into their equivalent string representations.

1. Create a new folder named `debugADC`, and within it, a new project. Set the RTE as we did for the previous recipe.

2. Create a new file, enter the following code, name the file `debugADC.c`, and add it to the project:

```c
/*---------------------------------------------------
 * Recipe:   debugADC_c2v0
 * Name:     debugADC.c
 * Purpose: Illustrates writing variables to GLCD
 *---------------------------------------------------
 *
 * Modification History
 * 06.02.14 Created
 * 09.12.15 Updated (uVision5.17 + DFP2.6.0)
 *
 * Dr Mark Fisher, CMP, UEA, Norwich, UK
 *--------------------------------------------------*/

#include "stm32f4xx_hal.h"
#include "GLCD_Config.h"
#include "Board_GLCD.h"
#include "Board_ADC.h"
#include <stdio.h>

#define wait_delay HAL_Delay

/* Globals */
extern GLCD_FONT     GLCD_Font_16x24;

#ifdef __RTX

/* Function Prototypes */
void SystemClock_Config(void);

/**
  * System Clock Configuration
  */
void SystemClock_Config(void) {

/**
```

```
     * Main function
     */
    int main (void) {
      char buffer[128];
      unsigned int ADCvalue;

      HAL_Init ( );
      SystemClock_Config ( );

      ADC_Initialize ();                    /* Initialse ADC */
      GLCD_Initialize ();                   /* Initialise GLCD */
      GLCD_SetBackgroundColor (GLCD_COLOR_WHITE);
      GLCD_ClearScreen ();
      GLCD_SetBackgroundColor (GLCD_COLOR_BLUE);
      GLCD_SetForegroundColor (GLCD_COLOR_WHITE);
      GLCD_SetFont (&GLCD_Font_16x24);
      GLCD_DrawString (0, 0*24, " CORTEX-M4 COOKBOOK ");
      GLCD_DrawString (0, 1*24, "        ADC Demo        ");
      GLCD_SetBackgroundColor (GLCD_COLOR_WHITE);
      GLCD_SetForegroundColor (GLCD_COLOR_BLUE);
      GLCD_DrawString (0, 3*24, "ADC =");

      for (;;) {        /* loop forever */
        ADC_StartConversion ();
        ADCvalue = ADC_GetValue ();         /* Read ADC */
        sprintf (buffer, "%i   ", ADCvalue);        /* mk str */
        GLCD_DrawString (7*16, 3*24, buffer);/* Disp it */
        wait_delay( 100 );
      } /* end for */
    }
```

3. Build, download, and run the program.

How it works...

The array named buffer just contains a collection of data elements, each the same type (in this case char). We need to specify the number of elements when the array is declared (so that the compiler can allocate the necessary storage space). This provides enough space for 128 characters.

> Strings are always terminated by a NULL character, so there is only space for 127 usable characters, but still plenty for our purpose.

The function `sprint()`, defined in the standard input/output C library that we've imported by using `#include <stdio.h>`, is used to convert the integer variable `ADCvalue()` to a string, placing the result in the buffer before being printed by `GLCD_DisplayString()`. The source code for the program is presented in step 2.

Running the program prints the 12-bit ADC value (generated by converting a voltage produced by the thumbwheel potentiometer) to the Graphic LCD display. Notice the values returned are quite noisy (that is, there is quite a bit of variation even when the thumbwheel position is apparently unchanged). If we shift the `ADCvalue` right by four places, using the bit manipulation operator `>>` so effectively discarding the least significant 4 bits (that is, dividing by 2^4), then the result is smaller and more stable.

Using the debugger

uVision5 provides a debugger that allows us to suspend execution (by inserting a breakpoint), and examine/change values of variables used in our program.

How to do it...

1. Download and run the previous project, `debug_ADC`.

2. Use the debug menu to insert a breakpoint on line 96 of our program (that is, at the statement `ADC_StartConversion ( );`.

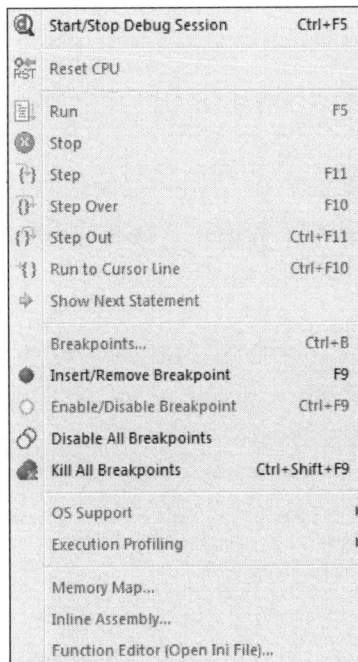

⊕	Start/Stop Debug Session	Ctrl+F5
RST	Reset CPU	
	Run	F5
⊗	Stop	
{↑}	Step	F11
{}→	Step Over	F10
{}↑	Step Out	Ctrl+F11
{}	Run to Cursor Line	Ctrl+F10
⇨	Show Next Statement	
	Breakpoints...	Ctrl+B
●	Insert/Remove Breakpoint	F9
○	Enable/Disable Breakpoint	Ctrl+F9
⊘	Disable All Breakpoints	
	Kill All Breakpoints	Ctrl+Shift+F9
	OS Support	▶
	Execution Profiling	▶
	Memory Map...	
	Inline Assembly...	
	Function Editor (Open Ini File)...	

3. Select **debug → Start/Stop Debug session** to start a debug session.

4. Observe that execution stops at main. This is because the default project debug options are set to **"Run to main"**.

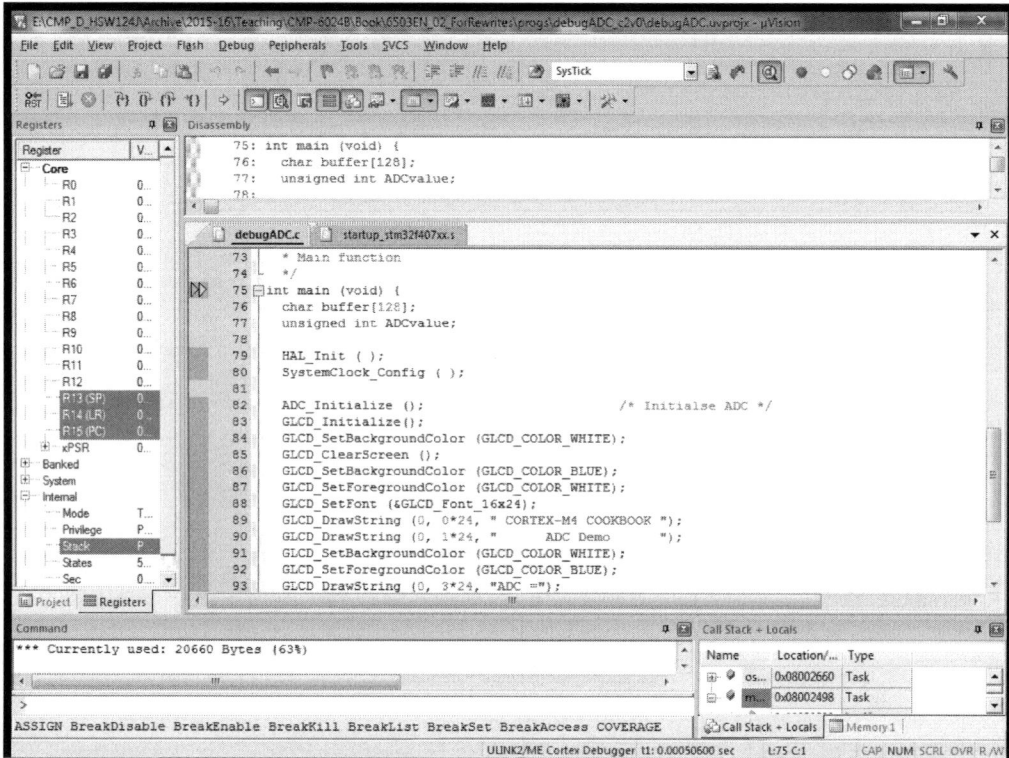

5. Selecting Run (*F5*) will execute the statements up until the breakpoint.

6. Use Step (*F11*) to execute the statements in the program one after the other, and observe the values of variables. For example, when we reach line 39 (after stepping), the local variable `ADCvalue` is assigned to 10 (*0x0000000A*). This value is shown in the **Call Stack + Locals** window.

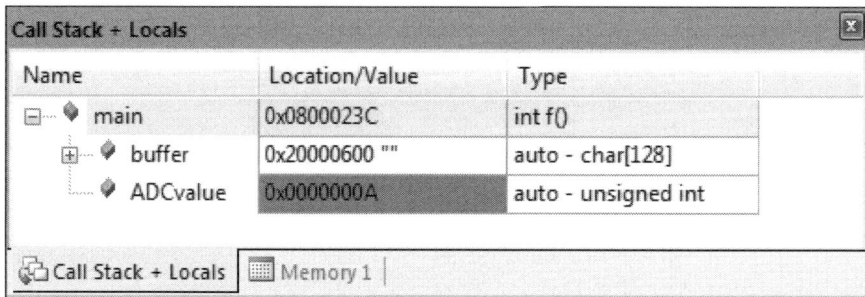

See also

This chapter has introduced many more programming concepts than would normally be covered in the first few chapters of a programming textbook, and the text is really aimed at those readers with experience of other languages. Those who are new to programming will need to fill some of the gaps by reading an introductory programming text. Because C has been around for more than 30 years, there are plenty to choose from! However, most novices will find recently published or revised editions of standard texts, more accessible than books written in the 1980s and 90s. You may find it easier to learn C by writing programs for your PC rather than the evaluation board. In fact, some of my students develop and test their embedded algorithms using a PC before porting them to uVision5. This is perfectly feasible for programs (or parts of programs) that do not need to access peripherals such as the ADC, and the like. You will need to install a C compiler to enable you to do this; free options include Visual Studio Express, Open Watcom, and GCC. The graphical user environments available with most of these compilers provide a user interface very similar to that of uVision5.

3

Programming I/O

In this chapter, we will cover the following topics:

- ▶ Performing arithmetic operations
- ▶ Illustrating machine storage classes
- ▶ Configuring GPIO ports
- ▶ Configuring UART ports
- ▶ Handling interrupts
- ▶ Using timers to create a digital clock

Introduction

The release of uVision5 heralded the integration of software packs to support a range of microcontroller devices and simplify the task of programming I/O by allowing the user to select from a menu of I/O options to provide the necessary source code in our project. This is extremely helpful and represents a huge leap forwards as compared to previous versions of the IDE that provided the user with comparatively little help with configuring I/O libraries. But, it does raise a dilemma; what do we do if our target hardware isn't supported? In this chapter, we'll investigate some of the functions that configure I/O devices and gain an understanding of what is involved in writing I/O interfaces for other targets. We'll need to refer to STM Reference manual *RM0090* (www.st.com) throughout this chapter as it provides complete information on how to use the STM32F405xx/07xx, STM32F415xx/17xx, STM32F42xxx, and STM32F43xxx microcontroller memory and peripherals. We start by writing a program that adds numbers and then use this apparently trivial code to motivate a deeper discussion of data types.

Performing arithmetic operations

Writing a program that adds two numbers together may seem like a trivial task. We obviously need to declare three variables, two to hold values of the numbers to be added, known as addends, and another to hold the sum. The following recipe illustrates some problems that arise due to word length.

How to do it...

The following steps demonstrate how to perform arithmetic operations:

1. Create a new folder and name it `addTwoNums_c3v0`. Invoke uVision5 and create a new project named `addTwoNums` within this folder.

2. Use the RTE manager to select the **MCBSTM32F400** evaluation board and configure it as we did for `helloWorld_c2v0`, from the *Writing to console Window* recipe in *Chapter 2, C Language Programming*.

3. Copy the files, `Serial.c`, `Serial.h`, and `Retarget.c`, from the `helloWorld_c2v0` recipe into the folder.

4. Create a new source file named `addTwoNums.c` and enter the following program. Please note that we're using the folding editor feature to omit boilerplate code:

```
/*****************************************************
 * Recipe:    addTwoNums_c3v0
 * File:      addTwoNums.c
 * Purpose:   Adds numbers using terminal I/O
 *****************************************************
 *
 * Modification History
 * 26.02.14 Created
 * 15.12.15 Updated uVision5.17 & DFP2.6.0
 *
 * Dr. Mark Fisher, CMP, UEA, Norwich, UK.
 *****************************************************/

#include "stm32F4xx_hal.h"
#include <stdio.h>
#include "Serial.h"
#include "cmsis_os.h"

#ifdef __RTX

/*-------------------------------------------------
   System Clock Configuration
```

```
 *------------------------------------------------------*/
void SystemClock_Config(void) {
```

```
/*
 * main
 *******/
int main (void) {

    int input;
    int num1, num2, res;

  HAL_Init ();     /* Init Hardware Abstraction Layer */
  SystemClock_Config ();            /* Config Clocks */

  SER_Init();

  for (;;) {                        /* Loop forever */
    printf("Enter First Number: ");
    scanf("%d", &input);
    num1 = (int) input;
    printf("Enter Second Number: ");
    scanf("%d", &input);
    num2 = (int) input;
    res = num1 + num2;
    printf("Result = %d \n", res);
  }
}
```

5. Add the `Serial.c`, `Retarget.c`, and `addTwoNums.c` files to the project.

6. Connect the evaluation board's UART 1/2/3 9-pin D-type connector to the PC's COM port.

7. Invoke PuTTY and configure the port as we did in *Chapter 2, C Language Programming*.

8. Check the **Use MicroLIB** project option.

9. Build the project; download and run the program (please note that you may need to reset the evaluation board).

10. Try adding a range of different values and make a note of the results (are they all correct?). Some examples are shown in the following screenshot:

```
COM1 - PuTTY
Enter First Number: 1
Enter Second Number: 2
Result = 3
Enter First Number: 10
Enter Second Number: 25
Result = 35
Enter First Number: 100
Enter Second Number: 155
Result = 255
Enter First Number: 100
Enter Second Number: 156
Result = 0
Enter First Number: 100
Enter Second Number: 200
Result = 44
Enter First Number:
```

11. Edit the main function and change the variable declaration for `num1`, `num2`, and `res` to the following:

    ```
    char num1, num2, res;
    ```

12. Rebuild, download, and run the code.

13. Try adding both positive and negative quantities and make a note of the results (are they all correct?). Try the examples that are shown in the following screenshot:

```
COM3 - PuTTY
Enter First Number: 1
Enter Second Number: 2
Result = 3
Enter First Number: 1
Enter Second Number: -2
Result = -1
Enter First Number: 100
Enter Second Number: 27
Result = 127
Enter First Number: 100
Enter Second Number: 28
Result = -128
Enter First Number: -100
Enter Second Number: -28
Result = -128
Enter First Number: -100
Enter Second Number: -29
Result = 127
Enter First Number:
```

How it works...

Programming languages classify the types of data they manipulate into categories called data types. Examples of data types are integer and floating point (numbers), character, string, and pointer. Simple (so-called primitive) data types are part of the language, while compound data types (such as array, struct, and so on) are abstractions built by the programmer. Programmers coding in strongly-typed languages (such as C) must declare the type of variables before they are referenced in the code. This enables the compiler to allocate a suitable amount of memory in which to store the variable. Typical primitive data types for the C language are shown in the following table. These types can be preceded by the signed or unsigned qualifier, which guarantees that the number is stored as a signed or unsigned quantity:

Type	Definition
char	This is the smallest addressable unit that can contain encoding of a character. It is, typically, 8-bits in size.
short short int	This is a short-signed integer type. It is *at least* 16-bits in size.
int	This is the basic signed integer type. This is at least 16-bits in size.
long long int	This is a long-signed integer type. It is at least 32-bits in size.

Type	Definition
long long long long int	This is a long-long signed integer type. It is at least 64-bits in size.
float	This is a single precision floating point type. Specific encoding is not specified, but IEEE 754 is a popular standard.
double	This is a double precision floating point type. Specific encoding is not specified, but IEEE 754 is a popular standard.
long double	This is an extended precision floating point type. Specific encoding is not specified, but IEEE 754 is a popular standard.

We've seen that unsigned numbers are stored in binary, but how are signed numbers represented? To answer this question, we'll consider the type char used to represent an 8-bit quantity. The type unsigned char encodes numbers between 0 and 2^8-1 that is illustrated as follows:

unsigned char								base$_{10}$
2^7	2^6	2^5	2^4	2^3	2^2	2^1	2^0	
0	0	0	0	0	0	0	0	0_{10}
0	0	0	0	0	0	0	1	1_{10}
.	.	.	.	.	.	.	.	.
1	1	1	1	1	1	1	1	255_{10}

The type definition also determines a set of valid operations on the type and how these are performed. For example, consider the arithmetic operation of addition. When we add two variables of type unsigned char, the result might be greater than 255_{10}. The rules of binary addition are illustrated in the following table:

SUM					CARRY
0	+	0	=	0	0
0	+	1	=	1	0
1	+	0	=	1	0
1	+	1	=	0	1

Each row in the table can be realized by digital hardware components (logic gates). When the sum is greater than 255_{10}, the result of the addition spills over into the eighth bit and gives the wrong answer. If we force the compiler to produce executable code for this operation, then the resulting operation would set the *CARRY* and *OVERFLOW* bits of the **Program Status Register** (**PSR**). The PSR forms a fundamental part of any central processing unit. Arithmetic instructions (and some others) that are executed by the CPU change the value stored in the five most significant bits of the STM32f4xx PSR register, setting or clearing them to reflect the outcome of the last arithmetic instruction that was executed:

31	30	29	28	27	28	0
N	Z	C	V	Q	Reserved	

- ▶ N=NEGATIVE
- ▶ Z=ZERO
- ▶ C=CARRY
- ▶ V=OVERFLOW
- ▶ Q=SATURATE

An operating system may read these bits and trap a run-time error. However, as our programs run without an operating system, and we've not included code to specifically trap errors, such operations may simply give the wrong answer when the data type that we're using is too small to represent the result. The following table illustrates adding 8-bit binary representations of 110_{10} and 198_{10}:

		2^7	2^6	2^5	2^4	2^3	2^2	2^1	2^0		10^2	10^1	10^0	
AUGEND		0	1	1 0 1			1	1	0	+	1	1	0	+
ADDEND		1	1	0	0	0	1	1	0		1	9	8	
SUM		0	0	1	1	0	0	0	0		3	0	8	
CARRY	1	1	0	0	1	1	1	0	Cin		1	0	Cin	

In this case, the 8-bit result overflows and is interpreted as 48_{10}.

Now consider the following assignment statement:

```
num1 = input;
```

Here the data types for num1 and input are declared as follows:

```
long int input;
unsigned char num1;
```

Remember that variable names are just pseudonyms for memory locations. The assignment statement copies the quantity stored in the memory location that is represented by the variable name on the right to the memory location that is represented by the variable on the left. But in this case, the problem is that these two are physically different sizes (that is, 8-bit and 32-bit, respectively). Typically, the compiler will report this as an error. To solve this problem, we must convert the 32-bit integer into 8-bit. The formal term for this is type conversion (also called type casting), and it is achieved using the following syntax:

```
num1 = (unsigned char) input;
```

If we wish to add both positive and negative quantities, we must change the data type of num1 and num2. Again, the range of numbers is limited by the size (number of bits) of memory used. If we use 8-bits to represent both positive and negative numbers, we must allocate half of the 256 binary codes to negative numbers and half to positive. Several systems have been proposed to achieve this (for example, signed magnitude, offset-binary, and 2's complement). The 2's complement system has four features that make its use in binary arithmetic very attractive. Firstly, the 8-bit code representing 0_{10} is 00000000_2. Secondly, negative values can be easily identified by examining the most significant bit (MSB). Thirdly, both positive and negative quantities can be added using the same simple logical operation that we identified, and finally, the algorithm to convert between positive and negative values is simply 'complement and add one'.

The char type is used to declare 8-bit numbers coded in 2's complement. Please note that using 8-bit 2's complement the largest positive number that can be represented is 2^7-1 (127_{10}) and the largest negative number -2^7 (-128_{10}).

Now consider the <stdio> library functions, scanf ( ) and printf ( ), that are used inside the superloop to establish a dialog with the user allowing them to enter values using the PC keyboard. Both functions use a so-called **format control string** to control the output and input format. A %d format string, is one of a number of integer conversion specifiers that are available to C programmers. The printf ( ) function uses %d to display signed decimal integers, and scanf ( ) uses it to read (optionally signed) decimal integers. Our program passes a pointer to the scanf ( ) function, so the long int variable named input is passed by reference and the function can change its value.

While working through the previous recipes you may have noticed that the type identifiers used in the Serial.h header file (supplied by Keil) are named differently from the primitive types that we encountered so far. The type identifiers, such as int32_t, and uint8_t, are called machine storage classes and represent pseudonyms for primitive types, such as int, and unsigned char. The next section discusses why we need them.

Illustrating machine storage classes

This recipe illustrates a version of addTwoNums that uses the machine storage classes, int32_t and uint8_t. We explain why it is advantageous for embedded applications to define and use these as opposed to the primitive types that are provided by the C language.

How to do it...

To define and use machine storage classes, please follow the outlined steps:

1. Create a new folder named addTwoNums_v2 by cloning the previous project.

2. Copy the addTwoNums.c file from the previous recipe to the folder and modify it as follows:

```c
int main (void) {

    int32_t input;
    uint8_t num1, num2, res;

    HAL_Init ();    /* Init Hardware Abstraction Layer */
    SystemClock_Config ();          /* Config Clocks */

    SER_Init ();

        for (;;) {                      /* Loop forever */
        printf ("Enter First Number: ");
        scanf ("%d", &input);
        num1 = (uint8_t) input;
        printf ("Enter Second Number: ");
        scanf ("%d", &input);
        num2 = (uint8_t) input;
        res = num1 + num2;
        printf ("Result = %d \n", res);
    }
}
```

3. Add the Serial.c, Serial.h, Retarget.c and addTwoNums.c files to the project.

4. Connect the evaluation board's UART 1/2/3 9-pin D-type connector to the PC's COM port.

Invoke PuTTY and configure the port as we did in *Chapter 2, C Language Programming*.

1. Remember to check the **Use MicroLIB** project option.

2. Build the project; download and run the program (please note that you may need to reset the evaluation board).

3. Check that the program behaves as before.

How it works...

The size of signed and unsigned integers that a microprocessor can manipulate is determined by its low-level architecture. The Cortex-M3 and -M4 microcontrollers are based on the ARMv7-M architecture (refer to *ARMv7-M Architecture Application Level Reference Manual*). Part A of the manual details the application-level architecture and programmers' model, and it begins by summarizing the core data types and arithmetic operations. ARMv7-M processors support the following data types in memory:

Byte	8-bit
Halfword	16-bit
Word	32-bit

The manual explains that processor registers are 32 bits in size, and the instruction set supports the following data types:

▸ 32-bit pointers

▸ Unsigned or signed 32-bit integers

▸ Unsigned 16-bit or 8-bit integers (held in zero-extended form)

▸ Signed 16-bit or 8-bit integers (held in sign-extended form)

▸ Unsigned or signed 64-bit integers held in two registers

It also describes the binary format that is used to store these quantities and provides a pseudo-code description of how addition and subtraction are performed. This description is consistent with the results that we got with the recipe, addTwoNums_c3_v0. The pseudo-code uses the terms *zero-extended* and *sign-extended* to describe how 8- and 16-bit numbers are stored in the 32-bit registers of the Cortex-M architecture. This is important as the processor status-register bits reflect the result of 32-bit arithmetic, and so, 8- and 16-bit values must be appropriately extended to fill the whole 32-bit register so that the sign and overflow bits correctly reflect the result of operations on shorter word lengths.

Implementations of C standard data types, such as char, short int, int, long int, and so on, depend on the (machine-specific) compiler implementation. You may recall that the C standard only specifies they must be at least a certain size. Apply italics to (at least). This can be a problem for embedded system programs that need to be ported between architectures with particular sizes of storage. Luckily, C provides a mechanism called `typedef` to create new types that are aliases of existing types. The C Standard Library includes `stdint.h`, containing C type definitions that can be customized for the different target architectures. The `stdint.h` header is included in `stm32F4xx_hal.h`, so there is no need to include it again in our program. A `typedef` keyword in the `stdint.h` header defines the following machine storage classes:

```
         /* exact-width signed integer types */
typedef    signed              char int8_t;
typedef    signed short        int int16_t;
typedef    signed              int int32_t;
typedef    signed            __int64 int64_t;

         /* exact-width unsigned integer types */
typedef unsigned               char uint8_t;
typedef unsigned short         int uint16_t;
typedef unsigned               int uint32_t;
typedef unsigned             int64 uint64_t;
```

If we require that an integer be represented in exactly N bits, then we use one of the following types:

signed:	int8_t	int16_t	int32_t	int64_t
unsigned:	uint8_t	uint16_t	uint32_t	uint64_t

Configuring GPIO ports

The recipe, `helloBlinky_c1v0`, that we met in *Chapter 1, A Practical Introduction to ARM Cortex*, uses the `LED_On()` and `LED_Off()` functions to switch the LEDs. These functions are defined in a file named `LED_MCBSTM32F400.c`, which is automatically included in our project if we select **LED (API) Board Support** when configuring our project using the RTE manager. Let's write another LED program and then take a closer look at `LED_MCBSTM32F400.c`.

How to do it...

To configure the GPIO ports follow the outlined steps:

1. Create a folder named `countBlinky_c3v0` and a project named `countBlinky`; use the RTE manager to select **Board Support** for the **LED (API)**.

2. Enter the following source code in file named `countBlinky.c` and add this to the project:

```c
/*-----------------------------------------------
 * Recipe:  countBlinky_c3v0
 * Name:    countBlinky.c
 * Purpose: LED Counter
 *-----------------------------------------------
 *
 * Modification History
 * 03.05.15 Created
 * 16.12.15 Updated (uVision5.17 + DFP2.6.0)
 *
 * Dr Mark Fisher, CMP, UEA, Norwich, UK
 *-----------------------------------------------*/

#include "stm32f4xx_hal.h"
#include "cmsis_os.h"
#include "Board_GLCD.h"
#include "Board_LED.h"

#define wait_delay HAL_Delay

#ifdef __RTX


/*-----------------------------------------------
  System Clock Configuration
 *-----------------------------------------------*/
void SystemClock_Config(void) {


/*
 * main
 ********/
int main (void) {
  uint8_t val = 0;

  HAL_Init ();      /* Init Hardware Abstraction Layer */
  SystemClock_Config ();            /* Config Clocks */

  LED_Initialize();           /* LED Initialization */

  for (;;) {                           /* Loop forever */
    LED_SetOut (val++);          /* increment LEDs */
    wait_delay(100);                     /* Wait */
  } /* end for */
}
```

3. Compile, download, and run the program.

How it works...

Computers access their I/O devices either by special I/O instructions that read and write to peripherals located in a separate I/O address space or using the instructions that are provided to access memory. ARM processors use the latter method, known as memory-mapped I/O. As such, peripheral registers are mapped into the memory address space of the machine, so turning LEDs ON and OFF is achieved simply by writing binary values to locations in memory.

As we explained in *Chapter 1, A Practical Introduction to ARM Cortex* each LED is connected to a GPIO port pin that in turn is mapped as a GPIO port bit. The GPIO interface is described in Reference manual *RM00090* (`www.st.com`), and it is impossible to understand the functions in `LED_MCBSTM32F400.c` without referring to this. The STM32F407IG has nine GPIO ports (named A-I), and each port can control up to 16 I/O bits. The port bits are configured as outputs or inputs by writing to so-called port control registers, and then data is either input or output by reading/writing to the data register that is associated with the port. Some port control bits configure programmable switches in the port that connect resistors to the pins. You may recall that LEDs need to be connected to resistors, so this feature is particularly useful. The switching speed of the port can also be configured by software (lower switching speeds save power). As ARM uses memory-mapped I/O, all GPIO registers are mapped to specific memory addresses.

Some evaluation boards connect all eight LEDs to one port, which makes configuring them easy, but the eight LEDs on the MCBSTM32F400 evaluation board are connected to different ports, and each port is dealt with separately. The `LED_On ()`, `LED_Off ()`, and `LED_SetOut ()` functions call `HAL_GPIO_WritePin ()`, which, in turn, is defined in the `stm32f4xx_hal_gpio.c` file. The GPIO registers themselves are declared as a C struct in the `stm32f407xx.h` file:

```
/**
  * @brief General Purpose I/O
  */

typedef struct
{
  __IO uint32_t MODER;      /*!< GPIO port mode register,
                                  Address offset: 0x00       */
  __IO uint32_t OTYPER;     /*!< GPIO port output type register,
                                  Address offset: 0x04       */
  __IO uint32_t OSPEEDR;    /*!< GPIO port output speed register,
                                  Address offset: 0x08       */
  __IO uint32_t PUPDR;      /*!< GPIO port pull-up/pull-down
                                  register,  Address offset: 0x0C       */
  __IO uint32_t IDR;        /*!< GPIO port input data register,
                                  Address offset: 0x10       */
  __IO uint32_t ODR;        /*!< GPIO port output data register,
```

```
                                   Address offset: 0x14       */
     __IO uint16_t BSRRL;     /*!< GPIO port bit set/reset low
                              register,  Address offset: 0x18       */
     __IO uint16_t BSRRH;     /*!< GPIO port bit set/reset high
                              register, Address offset: 0x1A        */
     __IO uint32_t LCKR;      /*!< GPIO port configuration lock
                              register, Address offset: 0x1C        */
     __IO uint32_t AFR[2];    /*!< GPIO alternate function registers,
                                   Address offset: 0x20-0x24 */
} GPIO_TypeDef;
```

In the C language, arrays and structures are compound data types used to store collections of data. All the data elements stored in an array must be the same size, (that is, all the same type), but in a **struct** (structure), the data values can be different sizes (types). As such, a struct provides an ideal abstraction for the data registers that are used by a peripheral. Each variable in the struct is accessed by a named identifier, which the compiler translates into an offset from a base address. In the previous example, the base address is represented by the GPIO_TypeDef identifier, and MODER, OTYPER, OSPEEDR, and so on represent offsets of 0, 4, 8, and so on bytes from the base (that is, *32 bits = 4 bytes*).

The typedef keyword enables the GPIO registers to be accessed using the GPIOx -> ODR syntax; for example, where GPIOx is a pointer to the base address of a particular GPIO port. Consider the HAL_GPIO_WritePin () function declared in Board_LED.h, which switches LEDs by writing to the bit-set-reset register (BSRR):

```
    void HAL_GPIO_WritePin(GPIO_TypeDef* GPIOx, uint16_t GPIO_Pin,

    GPIO_PinState PinState)
    {
      /* Check the parameters */
      assert_param(IS_GPIO_PIN(GPIO_Pin));
      assert_param(IS_GPIO_PIN_ACTION(PinState));

      if (PinState != GPIO_PIN_RESET)
      {
        GPIOx->BSRR = GPIO_Pin;
      }
      else
      {
        GPIOx->BSRR = (uint32_t)GPIO_Pin << 16;
      }
    }
```

Here, GPIOx is a pointer to the struct named GPIO_TypeDef that we described earlier. GPIOx->BSRRL writes '1' to a specific bit of the lower Bit Set Reset Register (BSRR) to set the port bit. BSRR controls bits 0-15 of the parallel port, as described in STM's *RM0090* Reference manual (*Chapter 8*) as follows:

31	30	29	28	27	26	25	24	23	22	21	20	19	18	17	16
BR 15	BR 14	BR 13	BR 12	BR 11	BR 10	BR9	BR8	BR7	BR6	BR5	BR4	BR3	BR2	BR1	BR0
w	w	w	w	w	w	w	w	w	w	w	w	w	w	w	w
15	14	13	12	11	10	9	8	7	6	5	4	3	2	1	0
BS 15	BS 14	BS 13	BS 12	BS 11	BS 10	BS9	BS8	BS7	BS6	BS5	BS4	BS3	BS2	BS1	BS0
w	w	w	w	w	w	w	w	w	w	w	w	w	w	w	w

Before we can use the GPIO to write to LEDs, the peripheral must first be configured. This is achieved by the LED_Initialize() function that is declared in Board_LED.h and defined in LED_MCBSTM32F400.c.

For example, within LED_Initialize(), the following code fragment configures GPIO Port G pins 6,7, and 8 to drive LEDs:

```
/* Configure GPIO pins: PG6 PG7 PG8 */
GPIO_InitStruct.Pin   = GPIO_PIN_6 | GPIO_PIN_7 | GPIO_PIN_8;
GPIO_InitStruct.Mode  = GPIO_MODE_OUTPUT_PP;
GPIO_InitStruct.Pull  = GPIO_PULLDOWN;
GPIO_InitStruct.Speed = GPIO_SPEED_LOW;
HAL_GPIO_Init(GPIOG, &GPIO_InitStruct);
```

For each port, LED_Initialize() writes appropriate values to a GPIO_InitStruct and then invokes HAL_GPIO_Init (). We need to consult STM's *RM0090* Reference manual yet again to fully understand HAL_GPIO_Init () (defined in stm32f4xx_hal_gpio.c), but some C language statements that are used by the functions read and write specific register bits are commonly used by embedded-system programmers and deserve further explanation. Consider this code fragment (in stm32f4xx_hal_gpio.c) that configures the GPIO Lock register (the function header provides a detailed description):

```
/**
  * @brief  Locks GPIO Pins configuration registers.
  * @note   The locked registers are GPIOx_MODER, GPIOx_OTYPER,
  *         GPIOx_OSPEEDR, GPIOx_PUPDR, GPIOx_AFRL and GPIOx_AFRH.
  * @note   The configuration of the locked GPIO pins can no
  *         longer be modified until the next reset.
  * @param  GPIOx: where x can be (A..F) to select the GPIO
  *         peripheral for STM32F4 family
  * @param  GPIO_Pin: specifies the port bit to be locked.
```

```
*            This parameter can be any combination of GPIO_PIN_x
*            where x can be (0..15).
* @retval None
*/


HAL_StatusTypeDef HAL_GPIO_LockPin(GPIO_TypeDef* GPIOx,
                                          uint16_t GPIO_Pin)
{
    __IO uint32_t tmp = GPIO_LCKR_LCKK;
    etc..
/* Apply lock key write sequence */
    tmp |= GPIO_Pin;
    /* Set LCKx bit(s): LCKK='1' + LCK[15-0] */
    GPIOx->LCKR = tmp;
```

The `tmp |= GPIO_Pin` statement assigns a value to `tmp`, which is a bitwise logical OR of the current value and a 32-bit mask named `GPIO_Pin`. The term mask is used to describe a binary variable that is used to identify particular bit patterns in a target variable. By carefully choosing the value of the mask, we are able to set particular bits of the **Lock Register** (**LCKR**) while maintaining the other bits unchanged. Please note that the `tmp |= GPIO_Pin` statement is written using a shorthand C assignment notation. To explain the notation, first consider a more familiar assignment such as the following:

```
myVar = myVar + 10;
```

This statement adds 10 to the variable myVar. This can be written in C shorthand as follows:

```
myVar += 10;
```

Another commonly used technique employs a bitwise logical AND operation with a mask to clear particular register bits. For example, `SER_Init ()` (recipe `addTwoNums_c3v0`) uses the following statement:

```
GPIOC->MODER &= 0xFF0FFFFF;
```

This is used to clear bits 20-23 of the GPIOC's MODE Register (`MODER`). Similarly, all operators can be combined in this way, so we could rewrite this as follows:

```
GPIOC->MODER &= ~(15UL << 20);
```

The ~ symbol represents the bitwise logical NOT operator, `15UL` is defined as an unsigned long of value 15, and `<<` is the logical shift-left operator.

Before explaining how the GPIO port's base address is defined, we'll deal with the type qualifier, `__IO` (refer to the `typedef` keyword that was illustrated earlier). The `__IO` macro is resolved by a `#define` directive in the `core_cm4.h` header file and replaced by the `volatile` qualifier. This qualifier indicates (to the compiler) that the variable is held in a register and may be changed by some external process. Typically, compilers optimize code by eliminating redundant loops that repeatedly read variables that are stored in memory. But, as we'll see in the next section, such busy-while loops are the key to many I/O operations, so the type `volatile` qualifier is essential when declaring I/O registers. Another commonly used qualifier is `__FORCE_INLINE`. This is used before a function definition to request the compiler to optimize the code by eliminating the function call.

The base addresses of GPIO ports are defined in the `stm32f407xx.h` file, as follows:

```
/*!< AHB1 peripherals */
#define GPIOA_BASE              (AHB1PERIPH_BASE + 0x0000)
#define GPIOB_BASE              (AHB1PERIPH_BASE + 0x0400)
#define GPIOC_BASE              (AHB1PERIPH_BASE + 0x0800)
#define GPIOD_BASE              (AHB1PERIPH_BASE + 0x0C00)
#define GPIOE_BASE              (AHB1PERIPH_BASE + 0x1000)
#define GPIOF_BASE              (AHB1PERIPH_BASE + 0x1400)
#define GPIOG_BASE              (AHB1PERIPH_BASE + 0x1800)
#define GPIOH_BASE              (AHB1PERIPH_BASE + 0x1C00)
#define GPIOI_BASE              (AHB1PERIPH_BASE + 0x2000)
```

Here, `AHB1PERIPH_BASE` is resolved by other `#define` statements and resolves to (uint32_t)0x40020000. This address is consistent with that identified in the *RM0090* Reference manual.

Peripherals are controlled by reading and writing to specific bits of the register bank and these are identified by so-called masks shown as follows (also defined in `stm32fxx.h`):

```
/********  Bits definition for GPIO_MODER register  ********/
#define GPIO_MODER_MODER0               ((uint32_t)0x00000003)
#define GPIO_MODER_MODER0_0             ((uint32_t)0x00000001)
#define GPIO_MODER_MODER0_1             ((uint32_t)0x00000002)

#define GPIO_MODER_MODER1               ((uint32_t)0x0000000C)
#define GPIO_MODER_MODER1_0             ((uint32_t)0x00000004)
#define GPIO_MODER_MODER1_1             ((uint32_t)0x00000008)
... etc.
```

The previous discussion illustrates the importance of the `stm32f407xx.h` header file. Take a moment to look through the source code. The comment at the beginning describes the content as "`CMSIS STM32F407xx Device Peripheral Access Layer Header File.`". Don't worry too much about the identifiers (such as `@file`, `@author`, `@version`, `@brief`, and so on). They are used by a tool to generate documentation from C (or C++) source code.

Finally, consider the following statement:

```
assert_param(IS_GPIO_PIN(GPIO_Pin));
```

This deserves some explanation. The `assert_param ()` macro is defined in the `stm32f4xx_hal_conf.c` file. A macro is defined as an instruction that expands to a set of instructions to perform a particular task. So, we would expect the following statement to appear somewhere in our project:

```
#define assert_param ... etc.
```

The macro definitions that we've met so far have been used to perform simple parameter substitutions, but `assert_param ()` introduces macro arguments, which makes macro behavior very similar to that of a function. If we take a look at the `assert_param` macro definition, we find the following:

```
/* Exported macro ---------------------------------*/
#ifdef   USE_FULL_ASSERT
/**
  * @brief   The assert_param macro is used for function's
  * parameters check.
  * @param   expr: If expr is false, it calls assert_failed
  * function which reports the name of the source file and
  * the source line number of the call that failed.
  * If expr is true, it returns no value.
  * @retval None
  */
#define assert_param(expr) ((expr) ? (void)0 :
                    assert_failed((uint8_t *)__FILE__, __LINE__))
/* Exported functions ------------------------- */
  void assert_failed(uint8_t* file, uint32_t line);
#else
  #define assert_param(expr) ((void)0)
#endif /* USE_FULL_ASSERT */
```

As the description explains, the macro checks that the `expr` input argument is TRUE, and if this is not the case, it calls `assert_failed ()`. It does this using a conditional statement that is written using C's only ?: ternary operator. Consider the following statement:

```
((expr) ? (void)0 : assert_failed((uint8_t *)__FILE__, __LINE__))
```

This statement is equivalent to the following:

```
If (expr)
   (void)0
else
   assert_failed((uint8_t *)__FILE__, __LINE__)
```

Defining this as a macro is more efficient as although it behaves as a function, the code is expanded by the preprocessor, and this avoids the overhead of an associated function call.

There's more...

Although memory mapped I/O is very efficient the memory address map is device and implementation is dependent, and this makes managing portability a problem. ARM solves this issue through the Cortex Microcontroller Software Interface Standard (CMSIS). CMSIS provides developers using the Cortex-M family with a common approach to interfacing peripherals, real-time operating systems, and middleware components. An overview of the standard `http://www.keil.com/support/man/docs` explains that it provides the following:

- A Hardware Abstraction Layer (HAL) for Cortex-M processor registers
- Standardized system exception names
- Standardized methods to organize header files
- Common methods for system initialization
- Standardized intrinsic functions
- Standardized ways to determine the system clock frequency

The following diagram shows CMSIS providing an interface between the user application (which may be based on a Real Time Operating System) and the hardware. CMSIS provides the following:

- A Core Peripheral Access Layer
- A Device Peripheral Access Layer (MCU-specific)
- Helper functions for peripheral management

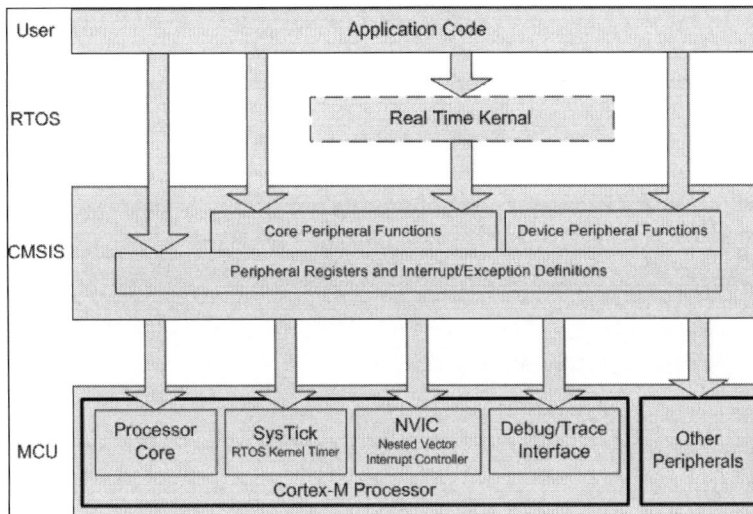

In practice, CMSIS is a framework within which MCU and peripheral vendors provide device driver libraries. Each vendor provides a device-specific {device} . h header file for users to include in their projects, and this may, in turn, include further files to provide additional functionality. MCU vendors also provide startup code written in assembly language that contains the vector table and initialization code for stacks, and so on. In the typical CMSIS file structure that is illustrated as follows, we see a number of file names that we are already familiar with through our previous projects:

- ▶ **{device}.h**: This is the header file defining the device
- ▶ **core_cm4.h**:This is the header file defining the device core
- ▶ **core_cm4.c**: This contains intrinsic functions
- ▶ **system_{device}.h**: This contains device-specific interrupt and peripheral register definitions
- ▶ **system_{device}.c**: This contains system functions and initialization code
- ▶ **startup_{device}.s**: This contains the startup code

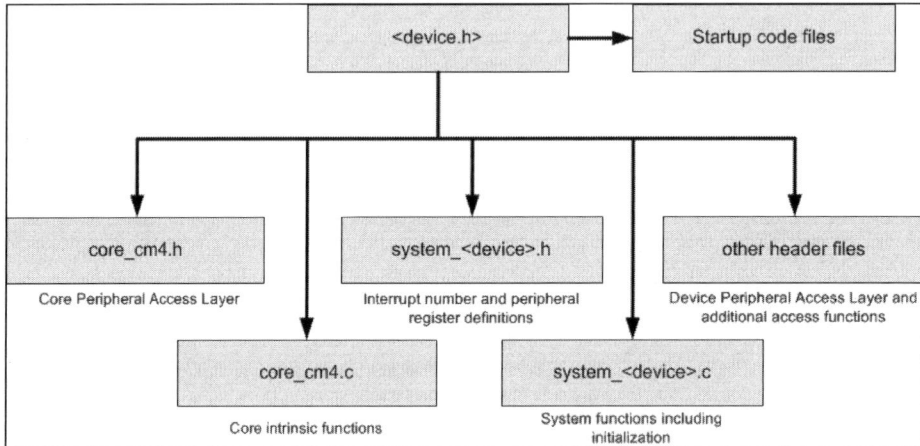

CMSIS continues to evolve as vendors develop new peripherals and revise how functionality is exposed by their device handlers. The version of CMSIS shipped with ARM's uVision4 IDE is quite different to the version that is shipped with uVision5, and judging by some of the comments posted on user forums, some users have found migrating to the new Run Time Environment manager quite a challenge. The main problem, especially for this text, is that some functionality has been packaged within the RTOS framework perhaps because this improves its robustness. More of a concern is that some of the functionality is only supported by the professional version of the MDK.

Configuring UART ports

Programs such as addTwoNums call the SER_GetChar() and SER_PutChar() functions to output ASCII characters to a terminal. The Retarget.c file redefines the fgetc() and fputc() functions, which, in turn, call SER_GetChar() and SER_PutChar(). These low-level functions illustrate some important I/O models that we'll explore using a program that checks if a string entered is a palindrome (for example, radar, civic, and level are palindromes). We'll call this recipe palindrome_c3v0.

How to do it...

Follow the steps outlined to configure UART ports:

1. Create a project named palindrome; use the RTE manager to configure the board as we did for addTwoNums_c3v0 folder, in the *Performing arithemetic operations* recipe.

2. Create a file named palindrome.c and copy the SystemClock_Config(void) function and associated boilerplate from a previous recipe. Add the following #include statements:

    ```
    #include "stm32F4xx_hal.h"
    #include <stdio.h>
    #include <string.h>
    #include "Serial.h"
    #include "cmsis_os.h"
    ```

3. Add a function named strRev () to the palindrome.c file:

    ```
    /*
     * strRev - returns reversed a string
     *******/
    char * strRev(char *str)
    {
        int i = strlen(str)-1,j=0;
        char ch;

        while(i > j)
        {
            ch = str[i];
            str[i]= str[j];
            str[j] = ch;
            i--;
            j++;
        }
        return str;
    }
    ```

4. Add a `main ()` function to the `palindrome.c` file and add this file to the project:

```
/*
 * main
 *******/
int main (void) {
  char a[100], b[100];

  HAL_Init ();    /* Init Hardware Abstraction Layer */
  SystemClock_Config ();           /* Config Clocks */

  SER_Init();

  for (;;) {
     printf("Enter the string to check for palindrome\n");
     scanf("%s", a);

     strcpy(b,a);
     strRev(b);

     if (strcmp(a,b) == 0)
       printf("Entered string IS a palindrome.\n");
     else
       printf("Entered string IS NOT a palindrome.\n");
  }
}
```

5. Remember to add `Retarget.c`, `Serial.c` and `Serial.h` to the project.

6. Open the project options dialog, click the **Target** tab, and check **Use MicrLIB**.

7. Build, download, and run the program.

How it works...

The a array stores the string that is entered. The string is copied to the b array and the `strrev()` function is called to reverse it. The `strcmp()` function (defined in the `string.h` library) is used to check whether the two strings match. The `strrev()` function copies and reverses the string character by character (remember that strings are terminated with a NULL character).

The `SER_PutChar()` function declared in `Serial.c` outputs characters by writing to the USART Data Register (DR), as follows:

```
/*----------------------------------------------------------------
 *        SER_PutChar:  Write a character to Serial Port
 *----------------------------------------------------------------*/
```

```
int32_t SER_PutChar (int32_t ch) {
#ifdef __DBG_ITM
  int i;
  ITM_SendChar (ch & 0xFF);
  for (i = 10000; i; i--);
#else
  while (!(UART4->SR & 0x0080));
  UART4->DR = (ch & 0xFF);
#endif

  return (ch);
}
```

The UART data register is referenced by a pointer:

```
UART4->DR;
```

> Please note that the STM32F4xx integrates both **Universal Synchronous/Asynchronous Receiver Transmitter** (**USART**) and **Universal Asynchronous Receiver Transmitter** (**UART**) hardware. USARTs can be configured to operate both synchronously and asynchronously. We configure a UART that is connected to the 9-pin D-type connector; hence, output is achieved by writing to UART4 rather than USARTx.

Once we have written to the data register, the digital value is output serially, one bit at a time, by the hardware. As this takes considerably longer than it takes to load data in parallel (the exact time taken will depend on the baud rate chosen), we must be careful not to load the DR with a new value until the previous one has been successfully transmitted. The previous line of code is as follows:

```
while ( !(UART4->SR & 0x0080) )
  /* empty statement */ ;
```

This line of code achieves this by checking bit 7 of the UART's Status Register (SR). Repeatedly reading the Status Register in a loop is called *polling* the Status Register (or *spinning on* the Status Register). A similar situation occurs in SER_GetChar (), but here we poll the Status Register to check whether a character has been received (that is, a bit-7 set), as follows:.

```
while ( !(UART4->SR & 0x0020) )
  ;
```

Polling or programmed I/O is the simplest I/O model that we can conceive and the corresponding empty while statements are known as busy-while loops. Programmed I/O operations are performed in the main thread of execution, so the busy-while loops prevent the CPU from doing any useful work. If the program is simple, then this is not too inconvenient, but in most cases, we must look to other more efficient I/O programming models, such as interrupt-driven I/O, and Direct Memory Access.

A flexible device driver really needs to support all three I/O models, that is, programmed I/O, interrupt-driven I/O, and DMA I/O. The USART device driver that is shipped with uVision 5 does exactly this. However, configuring this code is challenging, especially for novice programmers, so for the time being we'll develop our own simple drivers to gain some understanding of the mechanisms before migrating to ARM's library.

Embedded processors use serial ports to communicate with **Data Terminal Equipment** (**DTE**) and **Data Communications Equipment (DCE)** using the RS232 standard. Previous recipes, such as addTwoNums_c3v0, use RS232 to communicate with a PC running PuTTY to emulate a DTE. The RS232 signals are represented by voltage levels with respect to a system common (power / logic ground). The *idle* state (MARK) has the signal-level negative with respect to common, and the *active* state (SPACE) has the signal-level positive with respect to common. RS232 transmits data serially, as shown in the following figure:

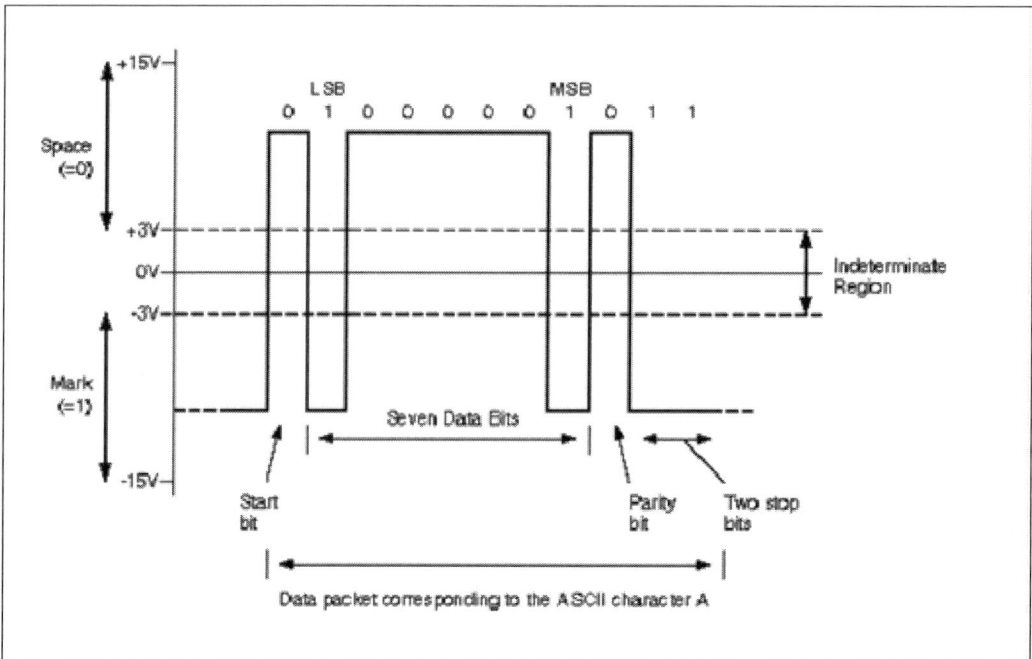

Data packet corresponding to the ASCII character A

Serial data is transmitted and received in normal USART mode as frames comprising the following:

▸ An Idle Line prior to transmission or reception

▸ A start bit

▸ A data word (7, 8, or 9 bits), the least significant bit first

▸ 0.5, 1, 1.5, or 2 stop bits, indicating that the frame is complete

The STM400Fxxx USART that is described in STM's Reference manual *RM00090* uses a fractional baud rate generator with a 12-bit mantissa and 4-bit fraction. The USART employs the following:

▸ A status register (USART_SR)

▸ Data Register (USART_DR)

▸ A baud rate register (USART_BRR)—12-bit mantissa and 4-bit fraction

▸ A Guardtime Register (USART_GTPR) in case of Smartcard mode

When data is transmitted asynchronously (without a shared common clock), the receiver and transmitter are synchronized by embedding timing information in the data signal by appending a "start" bit. The seven, eight, or nine data bits are appended after the start bit, a parity bit is added to detect errors, and the packet is terminated by one (or two) stop bits. The transmission rate (time allocated for each bit) is determined by the baud rate.

Configuring the USART involves writing appropriate values to the USART registers #ifdef and #else are preprocessor directives that facilitate conditional compilation):

```
/*-------------------------------------------------------------------
 *        SER_Init:  Initialize Serial Interface
 *-------------------------------------------------------------*/
void SER_Init (void) {
#ifdef __DBG_ITM  ITM_RxBuffer = ITM_RXBUFFER_EMPTY;
#else
  RCC->APB1ENR  |= (1UL << 19);            /* Enable USART4 clock */
  RCC->APB2ENR  |= (1UL <<  0);             /* Enable AFIO clock */
  RCC->AHB1ENR  |= (1UL <<  2);            /* Enable GPIOC clock */

  GPIOC->MODER   &= 0xFF0FFFFF;
  GPIOC->MODER   |= 0x00A00000;
  GPIOC->AFR[1]  |= 0x00008800;               /* PC10 UART4_Tx,
                                                 PC11 UART4_Rx (AF8) */

  /* Configure UART4: 115200 baud @ 42MHz,
                      8 bits, 1 stop bit, no parity */
  UART4->BRR = (22 << 4) | 12;
```

```
    UART4->CR3 = 0x0000;
    UART4->CR2 = 0x0000;
    UART4->CR1 = 0x200C;
#endif
}
```

Writing to the `UART4->BRR` baud rate register sets the baud rate. STM's Reference manual *RM00090* describes how to configure the Serial Ports. The baud rate is given

by $Tx/Rx\ baud = \frac{f_{clk}}{8(2\times OVERS)\times USARTDIV}$.

Where f_clk is the clock frequency of the USART clock, and `USART_DIV` is a 16-bit unsigned fixed-point number with a 12-bit mantissa and 4-bit fraction. In our case, we need a baud of 115200 and the APB1 domain clock is 48 MHz. Hence, missing f_clk again defined as eqn. object. $= 22.786_{10} = 0000000000010110.1100_2$, so DIV_Mantissa $= 22_{10}$, and DIV_Fraction $= 12/16$. Hence, referring to the description of the Baud Rate Register, we have the following:

```
    UART4->BRR = (22 << 4) | 12;
```

31	30	29	28	27	26	25	24	23	22	21	20	19	18	17	16
Reserved															
15	14	13	12	11	10	9	8	7	6	5	4	3	2	1	0
DIV_Mantissa(11:0)												DIV_Fraction(3:0)			
rw	rw	rw	rw	rw	rw	rw	rw	rw	rw	rw	rw	rw	rw	rw	rw

USART Control Register 1 provides some USART control functions:

31	30	29	28	27	26	25	24	23	22	21	20	19	18	17	16
							RESERVED								
15	14	13	12	11	10	9	8	7	6	5	4	3	2	1	0
OVER8	RESER-VED	UE	M	WAKE	PCE	PS	PEIE	TXEIE	TCIE	RXNEIE	IDLEIE	TE	RE	RWU	SBK
rw	res	rw	rw	rw	rw	rw	rw	rw	rw	rw	rw	rw	rw	rw	rw

Bits 2, 3, and 12 are set when 0x200C is written to Control Register 1 (CR1); this enables the USART (bit-12) and also enables the USART transmitter (bit-3) receiver (bit-2) functions. Bits 15, 12, and 9 are clear. This selects oversampling by 16 (bit-15), 8 data bits (bit-12), and even parity (bit-9). Bits 12:13 of CR2 are clear; hence, we have 1 stop-bit. Control register 3 functions are unused.

Other statements in `SER_Init()` connect appropriate clocks that are sourced from the **Real Time Clock Control** (**RCC**) peripheral and configure the GPIO to provide input and output for the USART by means of the Alternate Function logic. Please note that pins are an expensive microcontroller commodity, so GPIO pins are programmed to connect to a range of peripherals. We discuss GPIO Alternate Function in more detail in *Chapter 4, Programming I/O*.

Handling interrupts

This section illustrates an approach that improves on polling. We replace the busy-wait loop and instead configure the USART peripheral to generate an interrupt signal when a new character is received by the **input data register** (**IDR**). The interrupt signal causes a special function, known as an **interrupt service routine** (**ISR**), to be called, and this, in turn, reads the IDR and clears the interrupt signal. We illustrate this approach by a simple recipe called `helloISR_c3v0`.

Getting ready

Two small changes to `SER_Init()` are needed to configure UART4 so that interrupts are generated when a character is received. The value written to CR1 is changed from *0x200C* to *0x202C*, thereby setting bit-5 (RXNEIE), and the Nested Vectored Interrupt Controller (the NVIC is an ARM interrupt-dedicated peripheral close to the Cortex-M4 processor) is configured for UART4 as follows:

```
/*----------------------------------------------------------------
 *       SER_Init:  Initialize Serial Interface for interrupts
 *--------------------------------------------------------------*/
void SER_Init (void) {
  /* as before ... */

  /* Configure UART4: 115200 baud @ 42MHz,
                                    8 bits,
                         1 stop bit, no parity */
  UART4->BRR = (22 << 4) | 12;
  UART4->CR3 = 0x0000;
  UART4->CR2 = 0x0000;
  UART4->CR1 = 0x202C;

  /* Enable Interrrupts */
  NVIC_EnableIRQ(UART4_IRQn);
#endif
}
```

How to do it...

Follow these steps to handle interrupts.

1. Create a new folder (`helloISR_c3v0`) and within it a new project named `helloISR`; use the RTE manager to configure the project as we did for all the previous projects that use the serial port.

2. Create a file named `helloISR.c` and add the boilerplate code to configure clocks, and so on. Add this file to the project.

3. Add a function to handle interrupts from UART4, as follows:

```
/**********************************************************
 * UART4_IRQHandler
 *
 **********************************************************/
void UART4_IRQHandler (void) {
  volatile unsigned int IIR;
  volatile unsigned char c;

  IIR = UART4->SR;
  if (IIR & USART_FLAG_RXNE) { // read interrupt
    c = UART4->DR;
    printf("Interrupt! You pressed: %c \r\n", c);
    UART4->SR &= ~USART_FLAG_RXNE; // clear interrupt
  }
  else
    printf("Interrupt Error!\n");
}
```

4. Add a `main ()` function:

```
/*
 * main function
 ********/
int main (void) {

  HAL_Init ();    /* Init Hardware Abstraction Layer */
  SystemClock_Config ();           /* Config Clocks */

  SER_Init();
  printf ("Hello ISR I/O Example\r\n");
  printf ("Pressing a key generates an interupt\r\n");

  for (;;) {                       /* Loop forever */
    /* Nothing to do here */
  }
}
```

5. Remember to modify the `SER_init ()` function, as described previously.

6. Build, download, and run the program. Observe the response to keyboard strokes (illustrated in the next screenshot). Please note that when we test the code, it is best to configure PuTTY so that characters are not echoed to the terminal (as the ISR echoes the characters).

How it works...

Interrupts allow us to eliminate busy-while loops by providing a mechanism for the peripheral to initiate reads and writes to its I/O registers. It does this by sending a signal directly to the central processing unit (CPU) via the **Nested Vectored Interrupt Controller (NVIC).** This signal, called an interrupt, is automatically checked after each instruction is executed by the CPU, and, if active, the processor responds by executing a special function, known as an Interrupt Service Routine (ISR), that includes the read or write statement. Early processors were designed with only one interrupt signal, and several devices would be connected to this line using wired OR logic. In this case, when the interrupt occurred, the processor first needed to establish which device generated it before it could be serviced. The ARM Cortex employs a NVIC to manage up to 256 interrupts, each having a unique priority. This enables each device to call a unique ISR that is tailored to provide it with the service it needs. System events (for example, reset) and errors use exactly the same mechanism as interrupts but are called *exceptions* (to emphasize that they arise due to unusual system events). Both the interrupt and exception priorities are processor-specific and defined in `stm32F407xx.h`. The names of the ISRs are defined in the vector interrupt table, given in the `startup_stm32f407xx.s` file (the file extension, `.s`, indicates that this is an assembly language source file). Although interrupts solve the busy-while problem, they rely on the processor to read and write data to peripherals. While this is fine for a small number of data bytes, however, some peripherals (for example, Memory systems) handle blocks of data. So, we may find that a large chunk of the CPU time is consumed moving data rather than performing useful work. **Direct Memory Access** (**DMA**) solves this problem by enabling data to be moved directly between peripherals and memory. In this case, the data transfers are managed by a DMA controller, thereby leaving the CPU free to execute other more useful instructions.

Inspecting the interrupt vector table that is defined in `startup_stm32f4xx.s` allows us to identify the UART4 interrupt vector (that is `UART4_IRQHandler`). We must define a function named `UART_IRQHandler` to handle the interrupts. This ISR must read the USART status register (SR) and test the receive register not empty (RXNE) bit to confirm that the interrupt was generated by the port (if not, an error is indicated). Then the data register is read, echoed to the console terminal (PuTTY), and the interrupt is cleared (by writing zero to the RXNE).

The `SER_GetChar()` function in the `retarget.c` source file will need to be modified if we wish to use `stdio` library functions, such as `scanf()`, and so on. The best strategy would be to arrange for the ISR to write received characters to a buffer that could subsequently be read by `SER_GetChar()`.

There's more...

Interrupts provide a mechanism that allows the processor to multitask. Multitasking is a technique where a single processor divides its time between several instruction streams. This creates an illusion of parallelism as, to the user, it appears that different programs are executed concurrently when, in fact, they are not. Our programs that use ISR's have two threads of execution, but later we will write programs employing a real-time operating system kernel, and these may involve several threads. The differences between how normal threads and ISR threads are used have motivated processor designers to include features that enable multithreaded applications to be robust and recover from errors. Exceptions that are generated automatically when an error occurs are handled using exactly the same mechanism as interrupts and the term exception is generally used to describe either. When an error occurs, the strategy to recover from the exception may well involve reading/writing to processor registers that normal threads cannot access.

The Cortex-M4 processor operates in one of two modes. During the execution of the main program, the processor is in thread mode, and during execution of an exception handler or ISR, the processor is in handler mode. The two modes are distinguished by bits 0:8 of the PSR. In thread mode, bits 0:8 are zero, and in handler mode they are set to a number that identifies the exception type. As there are 8 bits, then 256 types of exceptions can be identified. When an exception is recognized the processor responds as follows:

1. The contents of processor registers R0:R3, R12, the return address, PSR, and link register (LR) are pushed to the active stack.

2. The processor identifies the exception number and uses this (offset) to access the interrupt vector table and locate the address of the exception handler, which is loaded into the program counter (PC).

3. The LR is loaded with a value that represents the execution mode of the processor (that is, thread or handler) prior to the exception having occurred.

4. The processor switches to handler mode and begins execution.

When the handler finishes, the return sequence pops the eight words from the stack and restores them to registers R0:R3, R12, LR, and PSR. It also loads the PC with the return address.

Access to special registers and system resources is determined by the privilege level of the processor. There are two levels, user and privileged. When in handler mode the processor is always in a privileged access level and can access all registers and memory resources. In thread mode, the processor is normally in user access privilege level and access to the System Control Space, an area of memory used to configure registers and debugging components, and access to some special registers is blocked. However, it is possible to switch from handler mode to user mode and maintain privileged access level, but the scenarios where this would be necessary are few. For most applications, the simple model of thread and handler modes that is shown as follows will suffice. After a reset, the processor is working in privilege mode in order to access all necessary resources.

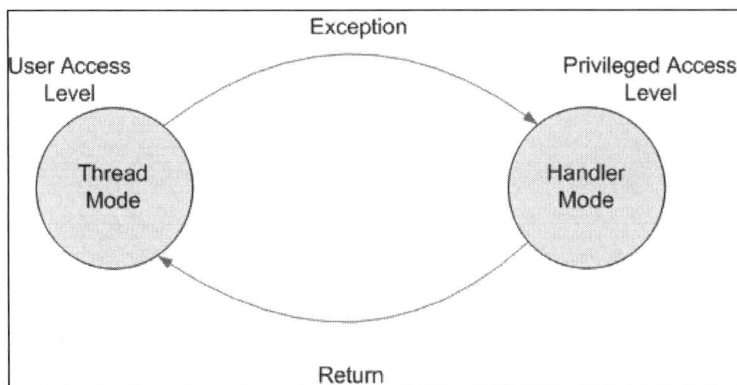

Using timers to create a digital clock

A digital clock application provides a good platform to illustrate the components that we discussed in this chapter. We'll use PuTTY to allow the user to set the time and then call HAL_GetTick () to provide a time-base for our digital clock that is displayed on the GLCD. We'll call this recipe ticToc_c3v0.

How to do it...

Follow the following steps to create a digital clock:

1. Create a new folder for the ticToc_c3v0 recipe and, within it, a new project (ticToc) and use the RTE manager to select board support for Graphic LCD.

2. Copy the retarget.c, serial.c and serial.h files to the project folder and add them to the project.

3. Define a new type (`time_t`) in the `ticToc.h` header file. Please note that we could declare each variable (hours, minutes, seconds, and so on) as separate unsigned integers, but it is better practice to group them together as a structured type named `time_t`:

```
#ifndef __TICTOC_H
#define __TICTOC_H

typedef struct {        /* structure of the clock record */
  unsigned char    hour;           /* hour */
  unsigned char    min;                       /* minute */
  unsigned char    sec;                       /* second */
} time_t;

#endif /*  __TICTOC_H  */
```

4. Create a new file named `ticToc.c`, add the necessary boilerplate and `#include` statements, and enter the following `main ()` function:

```
/*
 * main
 *******/
int main (void) {

  time_t time;
  int32_t input;
  char buffer[128];

  uint32_t tic, toc = 0;
  uint32_t elapsed_t;

  HAL_Init ();    /* Init Hardware Abstraction Layer */
  SystemClock_Config ();            /* Config Clocks */

  SER_Init();

  GLCD_Initialize();
  GLCD_SetBackgroundColor (GLCD_COLOR_WHITE);
  GLCD_ClearScreen ();            /* clear the GLCD */
  GLCD_SetBackgroundColor (GLCD_COLOR_BLUE);
  GLCD_SetForegroundColor (GLCD_COLOR_WHITE);
  GLCD_SetFont (&GLCD_Font_16x24);
  GLCD_DrawString (0, 0*24, " CORTEX-M4 COOKBOOK ");
  GLCD_DrawString (0, 1*24, "  PACKT Publishing  ");
  GLCD_SetBackgroundColor (GLCD_COLOR_WHITE);
```

```
GLCD_SetForegroundColor (GLCD_COLOR_BLACK);

for (;;) {                               /* Loop forever */

    }
}
```

5. Add `ticToc.c` to the project. Build, download, and test this. Please note that the compiler may issue some warnings as we have declared some unused variables.

6. Add the following code fragment immediately before the `for` statement:

```
/* Set the current time using PuTTY */
printf ("Clock Example\n");
printf ("Set Hours: ");
scanf("%d", &input); time.hour = input;
printf ("Set Minutes: ");
scanf("%d", &input); time.min = input;
printf ("Set Seconds: ");
scanf("%d", &input); time.sec = input;

/* elapsed_t is elapsed (10 * msec) since midnight */
elapsed_t =
        time.sec*100+time.min*60*100+time.hour*60*60*100;
```

7. Build, download, and test this.

8. Add the following code fragment within the `for` loop:

```
for (;;) {                               /* Loop forever */
  tic = HAL_GetTick()/10;
  if (tic != toc) {                      /* 10 ms update */
    toc = tic;
    time.sec = (elapsed_t/100)%60;    /* update time */
   time.min = (elapsed_t/6000)%60;
    time.hour = (elapsed_t/360000)%24;

    /* Update Display */
    sprintf(buffer, "%d : %d : %d", time.hour,
                            time.min, time.sec);
    GLCD_DrawString (4*16, 3*24, "              ");
    GLCD_DrawString (4*16, 3*24, buffer);

    elapsed_t = (elapsed_t+1)%DAY;
   }
  }
```

9. Remember to define the constant DAY as follows:

```
#define DAY 8640000;        /* 10 ms ticks in a day */
```

10. Compile, download, and run the program. The following below shows the GLCD screen:

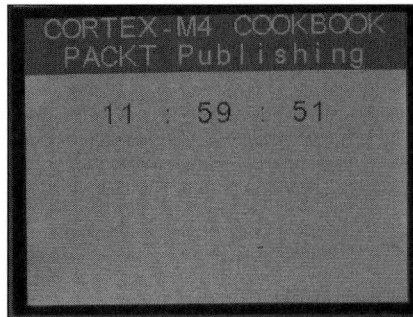

How it works...

Once we have declared a variable of type time_t, the fields (hours, min, sec) of the struct can be accessed using the dot operator (.) or the arrow operator (->). The dot operator accesses the structure field via the structure variable name, and the arrow operator accesses it via a pointer to the structure. We already used the arrow operator to access fields of structs that were used to represent peripheral registers. In this case, the arrow operator was used because the variable that was used to represent the struct (GPIOC, RCC, UART4 etc.) defines a pointer. As our main() function declares a variable (time) as time_t time;, we access the fields as time.hours, and so on.

The function named HAL_GetTick ( ) returns a value that is incremented every millisecond. We use this timebase to increment a counter variable named elapsed_t, which is initialized by the user's console (PuTTY). The tic and toc variables are updated to ensure that the display only needs to be updated every 100 ms. We call the function sprint() (declared in stdio.h) to format a string (stored in buffer [128]) representing the current time and write this to the Graphic LCD in a similar way to what was illustrated in this recipe, debugADC_c2v0.

4

Assembly Language Programming

In this chapter, we will cover the following:

- ▶ Writing Cortex-M4 assembly language
- ▶ Passing parameters between C and the assembly language
- ▶ Handling interrupts in assembly language
- ▶ Implementing a jump table
- ▶ Debugging assembly language

Introduction

This chapter explains how to write functions in assembly language. Assembly language is a low-level programming language that is specific to a particular computer architecture. So, unlike programs written in high-level languages, programs written in assembly language cannot be easily ported to other hardware architectures. Assembly language programs are converted into object code by a program called an assembler. In practice, assembly language is used only rarely and most embedded software is written in a high-level language, such as C. Assembly language is only used when the programmer needs precise control over the machine architecture and needs to access specific registers or when execution time is an important consideration. Such occasions typically occur during the following:

- ▶ Initializing the system
- ▶ Servicing I/O devices
- ▶ Handling interrupts

Assembly language programmers need a model of the computer architecture to enable them to write programs. This so called programmers' model identifies the registers, memory model, and instruction set for a particular machine architecture. The Cortex-M4 programmers' model is described in the *ARMv7-M Architecture Reference Manual* (please note that ARM restricts access to this document, but copies are available via third parties). *Chapter A2* identifies 13 general-purpose 32-bit registers (R0-R12) and three additional special registers (R13-R15) comprising Stack Pointer (SP), Link Register (LR), and Program Counter (PC). *Chapter A3* describes a flat address space that is 2^{32} bytes (4 GB) in size, and it identifies specific regions that are reserved for code, data, and memory-mapped I/O devices. The large number of instructions that are supported by the ARMv7-M architecture makes the prospect of writing an assembly language program quite daunting. A good strategy is to index instructions according to functionality (for example, branch, data processing, and data movement (load/store, and so on) as presented in *Chapter A4* of the reference manual.

The architecture supports a combination of 16-bit (Thumb) and 32-bit (ARM) instruction formats in an instruction set that is known as Thumb-2 technology. ARM's **Unified Assembler Language** (**UAL**) was developed to support both 16-bit and 32-bit instructions. We can specify a 32-bit instruction format using the .W (wide) suffix or the 16-bit format using the .N (narrow) suffix. If we omit the suffix, then the assembler chooses for us based on other constraints. For example, if the instruction references a high register (R8-R13), then a 32-bit variant must be used as most 16-bit instructions can only reference R0-R7. Instructions may also include other optional suffixes that identify the following:

▸ Status register flags in the Program Status Register (PSR) {cond} that determine execution (such as for branch instructions)

▸ If the condition flags are updated {S}

▸ The element size specified either as unsigned byte {B}, signed byte {SB}, unsigned half-word {H}, signed half-word {SH} or word (default)

The startup_stm32f407xx.s file that we include in all our projects is written in assembly language (conventionally, ARM file extensions, .s and .a, identify assembly language source code files). This is because one of its tasks is to set the Stack Pointer (SP) and writing to a specific register is impossible in C. Assembly language uses a mnemonic to represent each machine instruction. Other instructions called pseudo-instructions or assembler directives give commands to the assembler itself. Each line of the program combines instruction and pseudo-instruction mnemonics with operands and labels to carry out each program step. Labels may be included to act as symbolic references that are used to refer to memory locations, and so they save the programmer the tedious job of keeping track of addresses. An assembly language program is written so that labels, mnemonics, operands, and comments are arranged neatly in tabulated columns, that is,

```
[label]     [mnemonic]     [operand(s)]          ; [comment]
```

Each column must be separated by at least one whitespace character, and comments are preceded by a semicolon. Most assemblers are known as two pass because they parse the source code twice, first to build a table of symbolic references and associated addresses (called the symbol table) and again to produce the object code.

Writing Cortex-M4 assembly language

Before we start to write an assembly language subroutine, we need an idea of what the function has to achieve. The best way to specify this is to first write the function in a high-level language, such as C, and then translate the C code into assembly language line by line. Some experienced assembly language programmers argue that this is inefficient, but the technique produces well-documented code that can be optimized in further iterations of the design.

Getting ready

To translate the C code, we need to be familiar with both the instruction set and the addressing modes that are used by the Cortex-M4. Details of the instruction set can be found in ARM's Architecture Reference Manual and also in the ARM Cortex-M4 Generic User Guide (`http://infocenter.arm.com/`). Addressing modes are fundamental to a general understanding of computer architecture, but they are of practical interest to compiler writers and those writing assembly language. The following paragraph provides a very brief introduction.

The addressing mode describes the mechanism that an instruction uses to access its operands. In RISC architectures, such as the ARM Cortex, most instructions reference operands stored in registers directly (that is, register direct addressing). However, load and store instructions may reference a register value that is interpreted as an address in memory that contains the operand (that is, a pointer to the operand), so-called register indirect addressing. Additionally, if the value is interpreted as a pointer, then the effective address may be formed by adding an additional value (called the *offset*). The offset value may be specified as a constant and stored as part of the instruction, a technique known as *immediate addressing*, or stored in another register called an *index register*. The latter case, known as indexed addressing, is particularly efficient to access values stored in data structures, such as arrays, and structures. These addressing modes are summarized in the following table, and further information on additional addressing modes that are supported by the ARM Cortex-M4 can be found in Chapter A6 of the ARMv7-M Architecture Reference Manual.

Syntax	Addressing Mode	Description
<Rn>	Direct	This is the operand that is stored in the register
[<Rn>]	Indirect	This register holds a pointer to the operand

Syntax	Addressing Mode	Description
[<Rn>,<offset>]	Offset/Indexed Addressing	This is the effective address formed by adding the contents of base register <Rn> + <offset>. Offset may be the following: ▶ An immediate constant, for example, <imm8> or <imm12> ▶ An index register <Rm>

Consider translating the C code `const` declaration into assembly language, as follows:

```
const uint32_t delay = 10000;
```

ARM's Unified Assembler Language (described in the ARM compiler toolchain assembler reference `http://infocenter.arm.com/`) makes translating simple constant variable declarations very simple by providing a pseudo-instruction LDR that automatically produces appropriate ARM instructions to complete this task. In this case, assuming that we choose to store variable `num_ticks` in R0, then we can write the following:

```
;; Translating a const declaration
LDR R0, =10000    ; const uint32_t num_ticks = 10000;
```

The ARM assembler will convert this into an appropriate load instruction to move the required data value to the register. Let's suppose that we need to translate a variable declaration that doesn't include an assignment, as follows:

```
uint32_t cnt;
```

This doesn't require writing any code; we simply need to make a note of the register used to store the data:

```
;; Translating a variable declaration
;R1 <- cnt              ; uint32_t count;
```

We can then use LDR when a value is assigned:

```
;; Translating an assignment statement
LDR R1, =0        ; count = 0;
```

We will now illustrate the translation of a whole function by considering the simple delay used in the `helloBlinky_c2v2` recipe that was introduced in *Chapter 2, C Language Programming*. We'll call this recipe `asmBlinky_c4v0`.

How to do it...

1. Create a new project (in a new folder) called `asmBlinky`. Use the same RTE as `helloBlinky_c2v2` from the *Writing a function* recipe in *Chapter 2, C Language Programming*).

2. Make a copy of `helloBlinky.c` (the `helloBlinky_c2v2` folder from the *Writing a function* recipe in *Chapter 2, C Language Programming*.) and rename it `asmBlinky.c`.

3. Replace the comment at the start of the file with something more appropriate, let's take the following example:

```
/*-------------------------------------------------
 * Recipe:  asmBlinky_c4v0
 * Name:    asmBlinky.c
 * Purpose: Very Simple LED Flasher using
 *          Assembly Language delay function
 *-------------------------------------------------
 *
 * Modification History
 * 17.03.14 Created
 * 02.12.15 Updated
 * (uVision5 v5.17+STM32F4xx_DFP2.6.0)
 *
 * Dr Mark Fisher, CMP, UEA, Norwich, UK
 *-------------------------------------------------*/
```

4. Declare an external function called `delay ()`:

```
/* Function Prototype */
extern void delay(void);              /* asm subroutine */
```

5. Delete the C function named `delay ()` defined after `main ()` (a legacy of `helloBlinky.c`).

6. Add `asmBlinky.c` to the project.

7. Create a new file, enter the following assembly language code, and save the file as `delay.s`. Please note that the `.s` file extension is reserved for assembly language source code files:

```
;**********************************************************;
;* delay: Very simple assembly language delay routine *;
;*                                                    *;
;* Dr. Mark Fisher, CMP, UEA, Norwich, UK.            *;
;* Last updated 19.03.14                              *;
;**********************************************************;
        AREA    example, CODE, READONLY
        EXPORT delay      ;
delay                     ; void delay(void) {
```

```
      ;R0 <- num_ticks              ;     uint32_t num_ticks

              LDR R0, =10000000 ;
      ;R1 <- cnt                     ;     uint32_t cnt;
              LDR R1, =0            ;     for (cnt=0;
      cnt!=<num_ticks; cnt++)
      for     CMP R0, R1           ;
              BEQ forEnd           ;        /* empty statement */ ;
              ADD R1, #1           ;
              BAL for              ;
      forEnd                        ; }
              BX lr                ;
              END                  ;
```

8. Add `delay.s` to the project.

9. Build, download, and run the program.

How it works...

The name of the function translates to a label that acts as a pseudonym for the address of the start of the function. The variables are stored in R0 and R1 and assigned using LDR pseudo-instruction. R1 is incremented by adding an immediate constant (the immediate addressing mode is identified using #) to R1, the result is stored in R1. Its value is then compared to R0. The compare instruction subtracts R1 and R0 and sets the PSR flags but does not store the result of the operation. The conditional branch not equal (BNE) tests the zero flag and loads the program counter (PC) with the address of the label for if the flag is not set; otherwise, the program continues.

Programs often combine both C and assembly language functions, also known as subroutines. The assembly language code is written in a separate file that is read by the assembler. The main output produced by assembling an input assembly language source file is the translation of that file into an object file in **Executable and Linking Format** (**ELF**). ELF files produced by the assembler are relocatable files that hold code and/or data. The term relocatable means that all variables and branch targets are PC-relative, and so the code can be loaded anywhere in memory and executed. Relocatable ELF files produced by the assembler comprise of the following:

▶ An ELF header

▶ A Section header table

▶ Sections

Sections are the smallest independent, named, and indivisible units of code or data that are manipulated by the linker. The AREA assembler directive is used to subdivide our assembly language source file into ELF sections. Normally, we need at least two sections: one for program code, and another for data. There are two reasons for this. Firstly, some applications may store executable code in read-only memory (ROM), but variables need to be stored in read-write memory (RAM). Secondly, as the ARM Cortex-M4 uses a Harvard architecture model (that is, separate program and data memories) there is a considerable performance advantage in storing variables as data rather than code (even though both memories are implemented as nonvolatile RAM). As the examples we will investigate are not optimized for performance, our code and data can share the same section. Every program that includes assembly language must have at least one AREA directive (startup_stm32f407xx.s includes several).

As the delay assembly language subroutine is defined in another source file, then in order to call it from the main C function, we need to declare delay() as an external function. The name of the function resolves to the entry point in our assembly language subroutine (that is, an address), which is identified by adding the delay label in our code. We also need to include the EXPORT directive to enable the linker to resolve the symbol references.

When a function (written in C or assembler) is called, the program counter (PC) that holds the return address must first be saved and then overwritten with the address of the first instruction in the function. The ARM Cortex instruction set contains a primitive subroutine call instruction named **branch-with-link** (**BL**) that performs this function. When the function completes, a **branch indirect** (**BX**) instruction is used to load the PC register with the (saved) return address.

Every assembly language source file must end with an END assembler directive.

The ARM Architecture Procedure Call Standard (details in the next section) permits subroutines to overwrite R0-R3, so we can safely use R0 and R1 to store our local variables.

The AREA directive declares a CODE section called example that is READONLY and the delay label identifies the *ENTRY* to the subroutine. This symbol is exported to the linker. The R0 and R1 registers are used to hold the 32-bit const num_ticks and the cnt variable. Normally, one would need to save the contents of registers used by an assembly language subroutine; however, the ARM Architecture Procedure Call Standard (http://infocenter.arm.com/) permits subroutines to use R0-R3 without regard to their original contents (that is, their contents have been saved and restored by the calling function).

Values are loaded using the LDR pseudo-instruction and the register values are compared. If equal, the subroutine exits; otherwise, cnt is incremented. When the subroutine exits the register indirect branch, BX lr loads the PC register with the value given by the link register (R14).

There's more...

In addition to the object file identified by its file extension (.o), the assembler also creates a listing file (.lst) in the subdirectories named *Objects* and *Listings*. The listing file is very useful because it includes the instruction codes and the address labels used. A fragment of the listing for the `delay` subroutine is shown. This file can be a useful debugging aid. Please note that the comment field has been deleted for clarity:

```
 8 00000000                    AREA      example, CODE, READONLY
 9 00000000                                          ;
10 00000000                    EXPORT    delay       ;
11 00000000          delay                           ;
12 00000000          ;R0 <- num_ticks                ;
13 00000000 4804              LDR     R0, =10000000;
14 00000002          ;R1 <- cnt                      ;
15 00000002 F04F 0100         LDR     R1, =0       ;
16 00000006 4288    for       CMP     R0, R1       ;
17 00000008 D003              BEQ     forEnd       ;
18 0000000A F101 0101         ADD     R1, #1       ;
19 0000000E BFE8 E7F9         BAL     for          ;
20 00000012          forEnd                         ;
21 00000012 4770              BX      lr           ;
22 00000014                   END                  ;
```

See also

Documentation for the ARM Compiler Toolchain (including assembler directives) and ARM Instruction Set can be found in the **Tools Users' Guide** (accessed via uVision5's **Books** Tab).

Passing parameters between C and the assembly language

When a function or subroutine is called, its address must be loaded into the PC so that it can be executed and, when it terminates, execution of the calling routine must continue. In addition, there must be a convention that defines the following:

- ▶ How parameters are passed to the function
- ▶ How parameters are returned
- ▶ Which (if any) registers can be modified by the function

The ARM Architecture Procedure Call Standard deals with these issues (refer to Procedure Call Standard for the ARM Architecture, http://infocenter.arm.com/).

Getting ready

In this section, we will learn more about the **ARM Architecture Procedure Call Standard** (**AAPCS**) by writing an assembly language subroutine that receives a parameter from the C function that calls it. Moreover, in this example, the assembly language subroutine itself calls another C function. Functions or subroutines that call other functions or subroutines are called nested functions/subroutines.

How to do it...

We'll write another version of `helloWorld_c2v0` (introduced in the *Writing to the GLCD* recipe in *Chapter 2, C Language Programming*), but this time we'll use our own simple assembly language subroutine called `asmPrintf()`, instead of `printf()`, to output the string. We'll call this recipe `asmPrintf_c4v0`:

1. Create a new project (in a new folder) called `asmPrintf` by cloning `helloWorld_c2v0` (that is, use the same RTE as `helloWorld`).

2. Copy `helloWorld.c`; rename it `asmPrintf.c`.

3. Change the description to something more appropriate, as follows:

   ```
   /*-------------------------------------------------
    * Recipe:   asmPrintf_c4v0
    * Name:     asmPrintf.c
    * Purpose:  Outputs string using assembly language
    *           (illustrates parameter passing)
    *-------------------------------------------------
    *
    * Modification History
    * 23.03.14 Created
    * 17.12.15 Updated (uVision5 v5.17+DFP2.6.0)
    *
    * Dr Mark Fisher, CMP, UEA, Norwich, UK
    *-------------------------------------------------*/
   ```

4. Declare an external function named `asmPrintf ()`:

   ```
   /* function prototypes */

   extern void asmPrintf(char *);
   ```

5. Define a `main ()` function:

   ```
   /**
     * Main function
     */
   int main (void) {
   ```

```
      HAL_Init();
      SystemClock_Config();

      SER_Init();

      for (;;) {                          /* Loop forever */
        asmPrintf("Hello World!\n");
        wait_delay(1000);  }
    }
```

6. Add `asmPrintf.c` to the project.

7. Create a new file, enter the following assembly language code, and save the file as `asmPrintf.s`. Please note that the `.s` file extension is reserved for assembly language source code files:

```
;************************************************************;
;* A simple subroutine to print a string to the console *;
;************************************************************;
;*                                                        *;
;* Dr Mark Fisher, CMP, UEA, Norwich, UK                  *;
;* Last updated 23.03.14                                  *;
;************************************************************;
            AREA helloW, CODE, READONLY
            EXTERN SER_PutChar
            EXPORT asmPrintf
NULL    EQU 0                    ; #define NULL 0
asmPrintf                        ; void printf(char *ptr) {
            PUSH {R5, LR}        ;
; R5 <- ptr                      ;
; R0 <- c                        ;
            MOV R5, R0           ;
            LDRB R0, [R5], #1    ;    char c = *(ptr++);
while      CMP R0, #NULL        ;    while (c != NULL) {
            BEQ endWhl           ;
            BL SER_PutChar       ;      SER_PutChar(c);
            LDRB R0, [R5], #1    ;    char c = *(ptr++);
            B while              ;    }
endWhl     POP {R5, LR}         ;
            BX lr                ;
            END                  ; }
```

8. Add `asmPrintf.s` to the project.

9. Include `Retarget.c` and `Serial.c` in the project.

10. Connect the 9-Pin D-type UART1/3/4 connector on the evaluation board to the PC USB port (as we did in *Chapter 2, C Language Programming*).

11. Run the terminal emulator (PuTTy) configuring it as we did in *Chapter 2, C Language Programming*.

12. Build, download, and run the program.

How it works...

Our assembly language function needs a pointer to the first character of the string (exactly as `printf()` declared in `stdio.h` does), so our function prototype is as follows:

```
// Function prototype for assembly language subroutine
extern void asmPrintf(char *ptr);
```

As AAPCS uses `R0-R3` to hold the first four words of parameters passed to a function, we only need to pass one parameter (a pointer), so this is passed in `R0`.

As many novices find it difficult to write assembly language, we adopted the strategy of writing in C and then translating this code, line by line, into assembly language. The `asmPrintf()` C function is defined as follows:

```
// Function asmPrintf( )
void asmPrintf(char *ptr) {
  char c = *(ptr++);

  while (c != NULL) {
    SER_PutChar(c);
    c = *(ptr++);
  }
}
```

We include this in the comment field of our assembly language program to document the code. A key statement in the function is as follows:

```
c = *(ptr++);
```

This statement assigns a value to the c variable. The value is identified by a pointer variable, which is (later) incremented after the assignment is performed (so, `ptr` always points to the next character to be assigned to c). The while loop exits if c is a NULL character (strings are terminated by NULL characters).

The following is the assembly language instruction:

```
LDR{type} Rt, [Rn], #offset
```

This variant of LDR uses postindexed addressing; type determines the element size (that is, B, SB, H, and SH) and is omitted for word size memory access. Rt is the (target) register that we have to load. The address obtained from Rn is used as the address for the memory access. The offset value is added or subtracted from the address and written back to Rn.

To call the SER_PutChar() C function, the PC register must be loaded with its address. But as the function is defined in another file, we must leave it to the linker to sort out the detail. The EXTERN assembler directive identifies the SER_PutChar symbol as external. Working within the AAPCS, we must save any registers (other than R0-R3) that we use. When functions are nested then the link register (LR) must also be saved.

The ptr variable is passed in R0, but as R0 is needed to pass the input parameter to SER_PutChar(), we copy ptr to R5. The first instruction pushes the contents of R5 and LR onto the stack, and the last restores them, so the subroutine preserves state. Translating the while loop involves branching conditionally on the result of a comparison undertaken at the start of the loop.

There's more...

We can optimize the asmPrintf subroutine further using a Compare and Branch on Zero (CBZ) instruction. The instruction is as follows:

```
CBZ     Rn,label
```

This is equivalent to the following sequence:

```
CMP Rn,label
BEQ     label
```

However, Rn must be in the R0-R7 range, and the branch destination must be within 4-130 bytes of the instruction. Both of these restrictions are met in our case. A new version of our asmPrintf subroutine (asmPrintf_v2.s) is shown as follows:

```
;****************************************************************;
;* A subroutine to print a string to the console              *;
;****************************************************************;
;* Optimized using CBZ instruction (Cortex M3/M4)             *;
;*                                                            *;
;* Mark Fisher, CMP, UEA, Norwich, UK                         *;
;* Last Updated 26.03.14                                      *;
;****************************************************************;
        AREA helloW, CODE, READONLY
        EXTERN SER_PutChar
        EXPORT asmPrintf
```

```
asmPrintf                         ; void asmPrintf(char *ptr) {
            PUSH {R5, LR}         ;
; R5 <- ptr                       ;
; R0 <- c                         ;
            MOV R5, R0            ;    char c = *(ptr++);
            LDRB R0, [R5], #1     ;
while    CBZ R0, endWhl           ;    while (c != NULL) {
            BL SER_PutChar        ;        SER_PutChar(c);
            LDRB R0, [R5], #1     ;        char c = *(ptr++);
            B while               ;    }
endWhl      POP {R5, LR}          ;
            BX lr                 ;
            END                   ;    }
```

See also

In addition to the instruction set, an assembly language programmer also needs knowledge of the assembler directives, such as EQU, and so on. For further information, refer to the ARM Assembler Directives Reference (http://infocenter.arm.com/).

Handling interrupts in assembly language

ARM Cortex interrupt handlers can be programmed completely in C, but programmers coding time-critical applications prefer to use assembler (some programmers claim, rather ambitiously, that their hand-crafted assembler programs run up to 30-times faster than compiler generated code, but I suspect that the actual figure is 2-3 times).

When an interrupt (also known as an exception) occurs, the processor responds by performing the following actions:

▶ Pushing Registers R0-R3, R12, link register (LR), program counter (PC), and program status (PSR) onto the stack

▶ Reading the address of the exception handler from the interrupt vector table

▶ Updating the stack pointer, program status, link register, and program counter

The eight words pushed onto the stack are collectively known as the Stack Frame (illustrated later). These are referred to as caller-saved registers by the (AAPCS), and so the exception executes exactly as a C function. If the processor is in privileged mode, then the main stack will be used; otherwise the process stack is used.

The NVIC identifies the exception vector, and this is read from the vector table. On entry to the exception handler, either MSP or PSP is updated, the lower 8-bits of PSR (that is, ISR) are updated to show the exception number, the PC is loaded with the exception handler's address, and LR is loaded with a special value known as EXEC_RETURN, which is defined in the following table:

Bits 31:4	Bit 3	Bit 2	Bit 1	Bit 0
0xFFFFFFF	Return Mode	Return Stack	Reserved	Process State
	(thread/handler)		Must be 0	Thumb/ARM

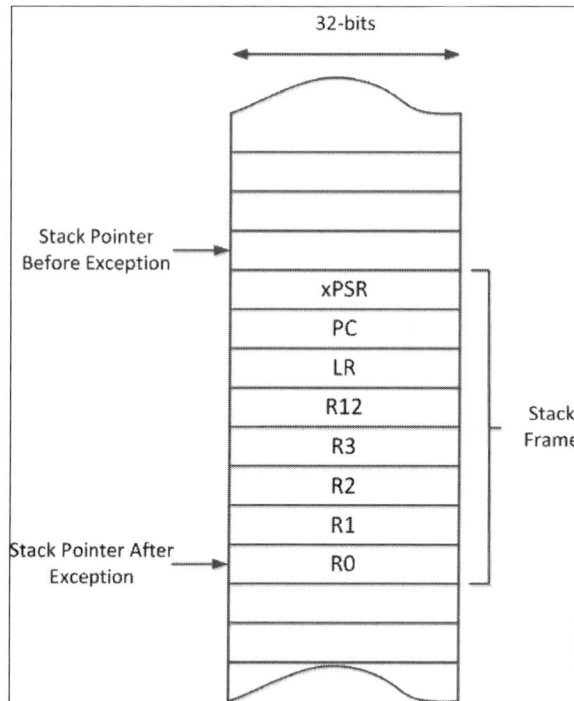

Stack Frame diagram showing 32-bits width. Stack Pointer Before Exception, followed by stack frame containing xPSR, PC, LR, R12, R3, R2, R1, R0. Stack Pointer After Exception points to R0.

Getting ready

To illustrate an assembly language interrupt handler, we'll translate the helloISR_c3v0 interrupt request handler recipe that we first introduced in the *Handling Interrupts* recipe in *Chapter 3*, *Programming I/O*. We call this recipe asmHelloISR_c4v0. An interrupt/exception handler must perform three steps:

- ▶ Process the interrupt request
- ▶ Deassert the request in the peripheral
- ▶ Return

How to do it...

1. Create a new project (in a new folder) called `asmHelloISR` by cloning `helloISR` (that is, use the same RTE as `helloISR_c3v0` introduced in *Chapter 3, Programming I/O*).

2. Copy the `helloISR.c` file and rename it `asmHelloISR.c`. Delete the C function named `UART4_IRQHandler ()` and add a new descriptive comment:

```
/*----------------------------------------------------
 * Recipe:   asmHelloISR_c4v0
 * Name:     asmHelloISR.c
 * Purpose:  Gets user key input using ISR
 *           (handler written in assembler)
 *----------------------------------------------------
 *
 * Modification History
 * 05.03.14 Created
 * 17.12.15 Updated
 * (uVision5 v5.17+STM32F4xx_DFP2.6.0)
 *
 * Dr Mark Fisher, CMP, UEA, Norwich, UK
 *----------------------------------------------------*/
```

3. Add this main function to the `asmHelloISR.c` file:

```
/**
 * Main function
 */
int main (void) {

  HAL_Init ();      /* Init Hardware Abstraction Layer */
  SystemClock_Config ();            /* Config Clocks */

  SER_Init ();

    printf ("Hello ISR I/O Example\r\n");
    printf ("Pressing a key generates an interupt\r\n");

  for (;;) {                        /* Loop forever */
        /* Nothing to do here */
  }
}
```

4. Add `asmHellISR.c` to the project.

5. Create a new file, enter the following code, and save it as `asmHelloISR.s`:

```
;*************************************************************;
;* Assembly language UART4_IRQHandler                      *;
;*************************************************************;
;* Dr Mark Fisher, UEA, Norwich, UK                        *;
;* Last Updated 26.03.14                                   *;
;*************************************************************;
          AREA example, CODE, READONLY
          EXPORT UART4_IRQHandler
          EXTERN printf        ;
UART4     EQU 0x40004C00       ;
SR        EQU 0x00             ;
DR        EQU 0x04             ;
RXNE      EQU 0x0020           ;
msg1      = "Interrupt! You Pressed: %c \r\n",0
msg2      = "Interrupt Error! \r\n",0
UART4_IRQHandler               ; void UART4_IRQHandler (void)
                               ; {
          PUSH {R4, LR}        ;
;R0 <- ptr                     ; char *ptr;
;R2 <- IIR                     ; unsigned int IIR;
;R1 <- c                       ; unsigned char c;
;R4 <- UART4                   ; uint32_t *UART4;
;
          LDR R4, =UART4       ;
          LDR R2, [R4, #SR]    ; IIR = UART4->SR;
if_       AND R2, #RXNE        ; if (IIR &
                               ;    USART_FLAG_RXNE) {
          CBZ R2, else_        ;
          LDR R1, [R4, #DR]    ;    c = UART4->DR;
          ADR R0, msg1         ;    ptr = msg1
;
          BL    printf         ;    printf(msg1, c);
          LDR R2, [R4, #SR]    ;    IIR = UART4->SR;
          AND R2, #~RXNE       ;    UART4->SR &=
          STR R2, [R4, #SR]    ;    ~USART_FLAG_RXNE;
          BAL ifend            ;    }
else_                          ;    else
          ADR R0, msg2         ;      printf("Interrupt
          BL    printf         ;             Error!\n");
ifend     POP {R4, LR}         ;
          BX    lr             ; }
          ALIGN
          END
```

6. Add `asmHelloISR.s` to the project.

7. Remember to add `Serial.c` and `Retarget.c` to the project.

8. Check **Use MicroLIB** in the project options dialog.

9. Connect the 9-Pin D-type UART1/3/4 connector on the evaluation board to the PC USB port (as we did in *Chapter 2, C Programming Language*).

10. Run the terminal emulator (PuTTY), configuring it as we did in *Chapter 2, C Programming Language*.

11. Build, download, and run the program.

How it works...

We need to write an assembly language subroutine called `UART4_IRQHandler` because this is the label referenced in the interrupt vector table that is defined in `startup_stm32f407xx.s`. As the handler must read and write to the registers of `UART4`, we also need its base address and the address offsets needed for the Status Register (`SR`) and Data Register (`DR`). This information can be found in the `stm32f407xx.h` header as follows:

```
;; UART4 addresses
UART4_BASE      EQU 0x40004C00      ; UART base address
SR              EQU 0x00            ; Status Register offset
DR              EQU 0x04            ; Data Register offset
```

We load `R4` with the `UART4` base address and use a *base + offset* addressing mode to load `R1`, the `UART` register. For example, the following sequence of instructions reads the Data Register:

```
LDR R4, =UART4_BASE        ;
LDR R1, [R4, #DR]          ; c = UART4->DR;
```

We also need to define masks to identify important flags, such as `SR` bit-5, and read data register not empty (`RXNE`):

```
RXNE         EQU 0x0020
```

We can define message strings to be output using the = pseudo instruction:

```
msg1         = "Interrupt! You Pressed: %c \r\n",&0
msg2         = "Interrupt Error! \r\n",&0
```

You will notice that C strings are automatically terminated by a `NULL` character, but in assembly language we need to explicitly tack `0` to the end. We'll use the `stdio` library's `printf()` function to output the string. This function takes two input arguments. The first is a pointer to the first character, and the second is the character argument referenced by the `%c` format specifier. We use the load PC-relative address assembly language instruction to load the location labelled as `msg1` into `R0`:

```
ADR R0, msg1
```

There's more...

Again, the ARM instruction set includes assembly language instructions that we can use to optimize things a little. The if-then condition instruction (IT) makes up to the four following instructions conditional. The conditions can be all the same or some can be the logical inverse of the others. The conditional instructions following the IT instruction are called the *IT* block. As there can be only four conditional instructions, we'll need to rewrite our C function so that it can be coded using an IT instruction:

```
Void UART4_IRQHandler (void) {
  uint8_t *ptr;
  uint32_t IIR;
  char c;
  uint32_t *USART_ptr;

  IIR = UART4->SR;
  c = (char) UART4->DR;
  if (IIR & USART_Flag_RXNE)
    printf("Interrupt! You pressed %c \r\n", c);
  else
    print("Interrupt Error!");
  USART4->SR &= ~USART_Flag_RXNE;
}
```

The changes that we made to `UART4_IRQHandler()` do not change its run time operation, but a compiler wouldn't be able to reorder the statements and ,thus take advantage of the if-then optimization. The complete subroutine is as follows:

```
;*************************************************************;
;* Assembly language UART4_IRQHandler                       *;
;*************************************************************;
;* Optimised using if-Then instruction                      *;
;*                                                          *;
;* Dr Mark Fisher, UEA, Norwich, UK                         *;
;* Last Updated 26.03.14                                    *;
;*************************************************************;
        AREA example, CODE, READONLY
        EXPORT UART4_IRQHandler ;
        EXTERN printf        ;
UART4   EQU 0x40004C00    ;
SR        EQU 0x00        ;
DR        EQU 0x04        ;
RXNE    EQU 0x0020        ;
msg1    = "Interrupt! You Pressed: %c \r\n",0
msg2    = "Interrupt Error! \r\n",0
```

```
        ALIGN
UART4_IRQHandler          ; void UART4_IRQHandler (void) {
        PUSH {R4, LR}
;R0 <- ptr                ;     char *ptr;
;R2 <- IIR                ;     unsigned int IIR;
;R1 <- c                  ;       unsigned char c;
;R4 <- UART4              ;     uint32_t *UART4;
;
        LDR R4, =UART4     ;
        LDR R2, [R4, #SR]     ;     IIR = UART4->SR;
        LDR R1, [R4, #DR]  ;     c = UART4->DR;
if_         AND R2, #RXNE  ;     if (IIR & USART_FLAG_RXNE) {
        CMP R2, #0         ;
        ITE NE             ;         printf("Interrupt! You
        ADRNE R0, msg1     ;             Pressed: %c \r\n"), c);
        ADREQ R0, msg2     ;     else
        BL    printf       ;         printf("Interrupt Error!\n");
        LDR R2, [R4, #SR]     ;
        AND R2, #~RXNE     ;   UART4->SR &= ~USART_FLAG_RXNE;
        STR R2, [R4, #SR]    ;
        POP {R4, LR}        ;
        BX    lr        ; }
        END
```

Implementing a jump table

Under certain circumstances, a jump table provides a very efficient way of implementing a
C language `switch` statement block. We can define a jump table as a list of unconditional
branch instructions—each referencing a different procedure or subroutine. We branch to one of
the subroutines by loading the program counter with the address of the unconditional branch
that is stored in the jump table. The effective addresses of items in the jump table are formed
using a base + offset addressing mode. Base + offset addressing is commonly used to access
data items stored in arrays, and a jump table is effectively just an array of address items.

Getting ready

To illustrate a jump table, we'll develop a recipe called `asmJumpTable_c4v0`. Assume that we
have a function named `jumpT ()` that accepts a `val` integer input argument. The function calls
either `proc1 ()`, `proc2 ()`, or `proc3 ()`, depending on the value of the input argument:

```
void jumpT ( int val ) {

  switch (val) {
    case 1 :
```

```
      proc1 ( );
      break;
    case 2 :
      proc2 ( );
      break;
    case 3 :
      proc3 ( );
      break;
    default :
      printf("Unrecognized!
                Enter value between 1-3\n");
      break;
  }
}
```

We'll implement `jumpT ()` in assembly language using a jump table.

How to do it...

1. Create a new project (in a new folder) called `asmJumpTable_c4v0` by cloning `asmHelloWorld` (that is, use the same RTE as `asmHelloWorld`).

2. Create a new file, enter the usual boilerplate, include the following, and save it as `asmJumpTable.c`:

```
#include "stm32F4xx_hal.h"
#include <stdio.h>
#include "Serial.h"
#include "cmsis_os.h"

/* Function Prototype */
extern void asmJumpT( int );
```

3. Add a main function, as follows:

```
/*
 * main
 *******/
int main (void) {

  int input, value;

  HAL_Init ();    /* Init Hardware Abstraction Layer */
  SystemClock_Config ();          /* Config Clocks */

  SER_Init ();
```

```c
    for (;;) {                              /* Loop forever */
      printf("\nJump Table Demo\n");
       printf("Enter Number from 1-3: ");
       scanf("%d", &input);
       value = (int) input;
      asmJumpT(value);
    }
}
```

4. Add `asmJumpTable.c` to the project.

5. Create a new file, enter the following assembly language code, and save the file as `asmJumpTable.s`. Please note that the `.s` file extension is reserved for assembly language source code files:

```
;*****************************************************;
;* A simple subroutine to illustrate a Jump Table    *;
;*****************************************************;
;*                                                   *;
;* Dr Mark Fisher, CMP, UEA, Norwich, UK             *;
;* Last updated 19.12.15                             *;
;*****************************************************;
;
        AREA example, CODE, READONLY
        EXPORT asmJumpT          ;
        EXTERN printf            ;
msg1    = "Case 1\n",0           ;
msg2    = "Case 2\n",0           ;
msg3    = "Case 3\n",0           ;
msgDef   = "Unrecognized! Value between 1-3 needed\n",0
                                 ;
        ALIGN                    ;
asmJumpT                         ; void JumpT(int val) {
        PUSH {R4, LR}            ;
        ADR r3, jumpTable        ;
; val->R0                        ;     switch (val) {
        SUB R0, #1               ;
        CMP R0, #2               ;
        BGT default              ;
        LDR pc, [r3,r0,LSL#2] ;      case '1' :
                                 ;          proc1( );
                                 ;          break;
                               ;        case '2' :
                                 ;          proc2( );
                                 ;          break;
                               ;        case '3' :
```

```
                                          ;            proc3 ( );
                                          ;            break;
        default                           ;          default :
                ADR R0, msgDef            ;            printf(msgDef);
                BL printf                 ;            break;
        endSW   POP {R4, LR}              ;        }
                BX lr                     ;    }
```

6. Add the jump table and associated subroutines to `asmJumpTable.s`:

```
jumpTable                                 ;
        DCD proc1                         ;
        DCD proc2                         ;
        DCD proc3                         ;
;**********************************************************;
;* Procedure 1                                          *;
;**********************************************************;
        ALIGN                      ; void proc1() {
proc1   ADR R0, msg1               ;
        BL printf                  ;   printf( msg1 );
        BAL endSW                  ; }
;**********************************************************;
;* Procedure 2                                          *;
;**********************************************************;
        ALIGN                      ; void proc2() {
proc2   ADR R0, msg2               ;
        BL printf                  ;   printf( msg2 );
        BAL endSW                  ; }
;**********************************************************;
;* Procedure 3                                          *;
;**********************************************************;
        ALIGN                      ; void proc3() {
proc3   ADR R0, msg3               ;
        BL printf                  ;   printf( msg3 );
        BAL endSW                  ; }
                                   ;
        END                        ;
```

7. Add `asmJumpTable.s` to the project.

8. Remember to add `Serial.c` and `Retarget.c` to the project.

9. Connect the 9-Pin D-type UART1/3/4 connector on the evaluation board to the PC USB port (as we did in *Chapter 2, C Programming Language.*).

10. Run the terminal emulator (PuTTY), configuring it as we did in *Chapter 2, C Programming Language.*

11. Compile, download, and run the program.

How it works...

The jump table is defined as follows:

```
jumpTable                    ;
        DCD proc1            ;
        DCD proc2            ;
        DCD proc3            ;
```

Here, `proc1`, `proc2`, and `proc3` are address labels that are used to identify the start of the subroutines. The `jumpTable` base address is loaded into R3 by the ADR pseudo-instruction:

```
        ADR r3, jumpTable        ;
```

The assembler attempts to replace ADR to produce a single ADD or SUB instruction to load the address using a PC-relative addressing mode. This ensures that ADR always assembles to one instruction. The assembler will produce an error if it can't load the effective address in one instruction. The most likely reason for this will be that the target base address is too far away, and we will need to replace ADR with ADRL.

The value passed in R0 will be an integer between 1-3, so subtracting 1 will give the address offset directly:

```
        SUB R0, #1                ;
```

Finally, we use the following to load the program counter with the appropriate jump table address (that is, entry 1, 2, or 3):

```
        LDR pc, [r3,r0,LSL#2] ;
```

Each jump table entry is a 32-bit (4-byte) address, so the value in R0 needs to be multiplied by 4 (that is, LSL #2). This is achieved by LDR, and the Register Offset instruction. Finally, a Branch and Link instruction BL is needed to execute the function.

We've used the ALIGN pseudo-operation quite liberally in all our assembly language programs. ARM compilers normally access data in memory aligned on word boundaries and pad data structures so that items can be accessed efficiently. Consequently, address labels need to be placed on word boundaries. The ALIGN pseudo-operation ensures this. Leaving it out will produce a message from the assembler warning that some padding has been inserted.

Debugging assembly language

We can gain a useful insight into how assembly language instructions execute, and also why the compiler is rather poor at translating C using the debugger.

First, we'll compare a fragment of assembly language code produced by the compiler with our translation.

How to do it...

1. Open the `helloISR_c3v0` recipe that we introduced in the *Handling interrupts* recipe *Chapter 3, C Language Programming*.

2. Insert a breakpoint adjacent to the first statement of the `UART4_IRQHandler` (that is, `IIR = UART4->SR;`).

3. Select **Debug → Start/Stop Debug Session** from the uVision5 pull-down menu.

4. Run (*F5*) to the breakpoint (you will need to select the console window (PuTTY) and enter a character).

5. uVision5 will now open a **Disassembly** window (illustrated in the following screenshot), which shows the assembly and machine code generated by the compiler for each C language statement.

```
Disassembly                                                        ⏸ ☒
       27: void UART4_IRQHandler (void) {
       28:          volatile unsigned int IIR;
       29:          volatile unsigned char c;
       30:
0x08000230 B51C       PUSH           {r2-r4,lr}
       31:          IIR = UART4->SR;
0x08000232 4813       LDR            r0,[pc,#76]  ; @0x08000280
0x08000234 8800       LDRH           r0,[r0,#0x00]
0x08000236 9001       STR            r0,[sp,#0x04]
       32:          if (IIR & USART_FLAG_RXNE) { // read interrupt
0x08000238 9801       LDR            r0,[sp,#0x04]
0x0800023A F0100F20   TST            r0,#0x20
0x0800023E D010       BEQ            0x08000262
       33:              c = UART4->DR;
0x08000240 480F       LDR            r0,[pc,#60]  ; @0x08000280
0x08000242 1D00       ADDS           r0,r0,#4
0x08000244 8800       LDRH           r0,[r0,#0x00]
0x08000246 B2C0       UXTB           r0,r0
0x08000248 9000       STR            r0,[sp,#0x00]
```

How it works...

Some interesting observations from the disassembly are evident. First, by default, the compiler stores its variables in memory (rather than registers), so assignment statements resolve to a sequence of load (`LDR`) and store (`STR`) instructions. Overall, the compiler produces slightly more assembly language instructions than an assembly language programmer coding by hand.

There's more...

Now, open `asmHelloISR_c4v0`, which was introduced in the *Handling interrupts in assembly language* recipe:

1. Place a breakpoint at the first instruction of the assembly language subroutine `UART4_IRQHandler` (make sure you identify an ARM instruction and not a label or pseudo instruction).

2. Use the debugger to run to the breakpoint, as illustrated in the following screenshot. Now, use the step (*F11*) command and observe the register contents changing as each instruction is executed:

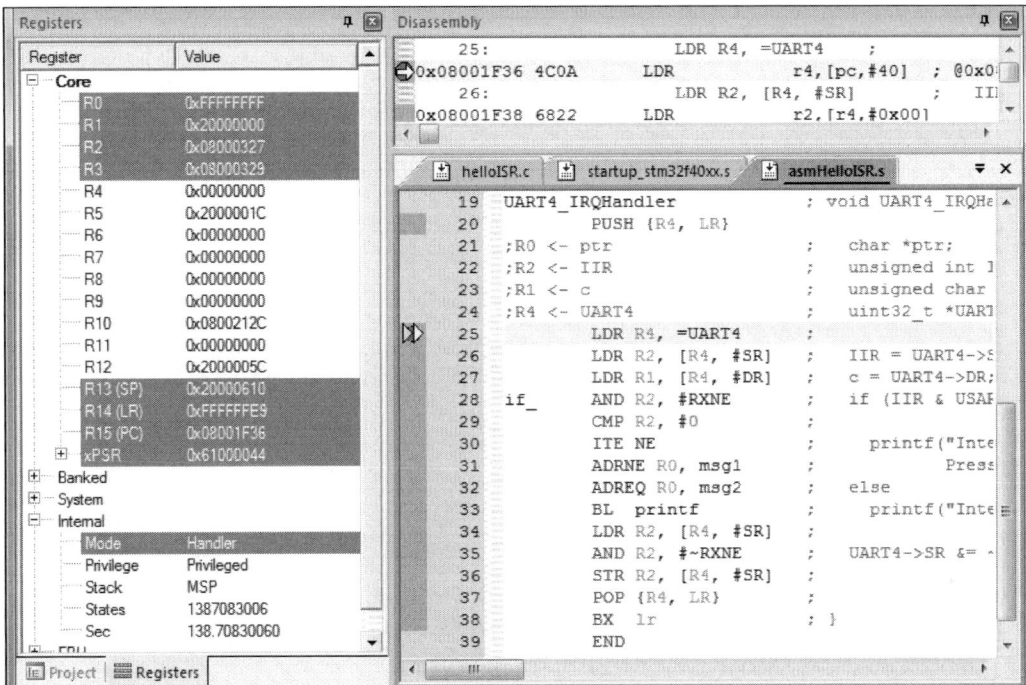

You will notice that observing how register values change as we single step through assembly language code provides a useful insight into the operation of the Cortex-M4 machine architecture.

5
Data Conversion

In this chapter, we will cover the following topics:

- ► Setting up the ADC
- ► Configuring general-purpose timers
- ► Using timers to trigger conversions
- ► Setting up the DAC
- ► Generating a sine wave

Introduction

Most signals that we encounter in the natural world are continuous; for example, we perceive sound produced by an orchestra as a continuum of intensities ranging from *pianissimo* (very soft) to *fortissimo* (very loud). Computers, on the other hand, work with binary quantities that are inherently discrete. The number of discrete values that can be represented depends on the number of bits that are used to represent the quantity (for example, 8 bits can represent 2^8 discrete values). Computers that are designed to interact with real-world phenomena (for example, sound, light, heat, and so on) need to overcome two problems. Firstly, they need to convert between its physical manifestation and a (continuous) electrical signal, and secondly, they need to convert between the signal's continuous and discrete representation. Returning to our sound example, solving the first problem requires a transducer to convert sound (pressure) waves to electrical signals and vice versa (that is, a microphone and loud speaker). Solving the second requires converting the analog (continuous) signal to a discrete form and vice versa. The device that is used to achieve this is called an **Analog-to-Digital converter** (**ADC**)—and conversely a **Digital-to-Analog converter** (**DAC**).

Analog-to-Digital conversion requires measuring (sampling) the signal at regular time intervals and converting each sample into a digital value. This raises the question, how often should we take the measurement? This fundamental question is addressed by signal processing theory. The short answer is that samples must be taken at least twice as frequently as the period of the highest-frequency component in the signal. However, the maximum number of samples that can be taken every second (that is, the maximum sampling frequency) is limited by the speed of conversion, and this, in turn, depends on the type of ADC. The STM32F407IG microcontroller includes a successive approximation ADC, which is fast enough for most audio applications (that is, signals having frequency components up to about 20 KHz). A block diagram of a successive approximation ADC is shown as follows:

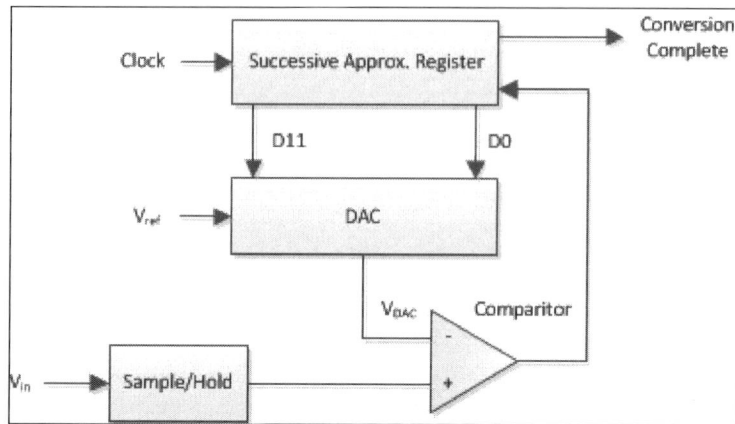

Successive approximation ADC

A single comparator is at the heart of the successive approximation ADC. This is simply a device that outputs a binary signal that depends on a comparison of V_{DAC} and V_{in}, where V_{DAC} represents the analog voltage corresponding to the output of the **Successive Approximation Register** (**SAR**). By testing the output of the comparator, an algorithm aims to update the SAR so as to find the value V_{DAC} that is closest to V_{in}. The successive approximation DAC achieves this by undertaking a search that aims to find V_{DAC} ($\leq V_{ref}$) in the fewest number of guesses. The time needed for the search depends on the value of the voltage, but the worst-case conversion time ultimately determines the maximum sampling frequency. The DAC is a much simpler analog circuit that uses a summing amplifier to add together the (weighted) digital outputs D0-D11. Hence, the DAC operates much faster than the ADC.

The purpose of the **Sample/Hold** block is to take a snapshot of the input voltage, and so, provide a stable signal for the ADC. The Sample/Hold block is not ideal and it takes some time (called the aperture time) to capture the input signal. The signal voltage stored by the Sample/Hold block also decays with time, but the Sample/Hold time can be adjusted to address these problems. A range of values can be specified in terms of a number of ADC clock cycles by writing to the two ADC **sample time registers** (SMPR1 and SMPR2). The time can be set for each channel using the following codes:

SMPx[2:0]	Chan.x Sample Time (cycles)	SMPx[2:0]	Chan.x Sample Time (cycles)
000	3	100	84
001	15	101	112
010	28	110	144
011	56	111	480

The maximum conversion time, T_{conv}, for a successive approximation converter is equal to the *Sample/Hold time + (clock period × number of bits)*. As a rule of thumb, it's best to make the Sample/Hold time short relative to the sample period.

Setting up the ADC

The aim of this recipe is to configure the ADC in single-conversion mode and then convert the voltage set by the thumbwheel into a 12-bit digital value. We'll configure the ADC to generate an interrupt at the end of each conversion and write an interrupt handler to read the ADC and initiate a new conversion. The only task for our main function to perform is to output the ADC value to the LEDs, but as there are only 8 LEDs we can only display the most-significant 8-bits of the ADC value. We'll call this recipe adcISR_c5v0.

How to do it...

To set up the ADC follow the steps outlined:

1. Open a new folder named adcISR_c5v0 and create a new project named adcISR. uvprojx.

2. Select **LED (API)** from RTE **Board Support** but do not select **A/D converter** (we will write our own code for this). Set the **CMSIS** and **Device** software components as for previous projects. Be sure to select **resolve** so that the correct runtime environment (RTE) is included.

3. Create an `adcISR.c` file (the main function) and enter the source code that is shown next. Remember to include the boilerplate code (hidden by the editor folds):

```
/*-----------------------------------------------------
 * Recipe:  adcISR_c5v0
 * Name:    adcISR.c
 * Purpose: A/D Conversion Demo for MCBSTM32F400
 *          using IRQ
 *-----------------------------------------------------
 * Modification History
 * 16.04.14 created
 * 22.12.15 updated uVision5.17 + DFP2.6.0
 *
 * Dr Mark Fisher, UEA, Norwich
 *-----------------------------------------------------*/

#include ""stm32f4xx_hal.h""
#include ""Board_LED.h""
#include ""Custom_ADC.h""

#define wait_delay HAL_Delay

/* Globals */
uint32_t adcValue;

#ifdef __RTX
```

```
/* Function Prototypes */
void SystemClock_Config(void);

/**
  * System Clock Configuration
  */
void SystemClock_Config(void) {
```

4. Include code to handle the interrupt generated by the ADC:

```
void ADC_IRQHandler (void) {

    ADC3->SR &= ~2;          /* Clear EOC interrupt flag */
    adcValue = (ADC3->DR);      /* Get converted value */
    ADC3->CR2 |= (1 << 30);  /* Start next conversion */

}
```

5. Include a `main ()` function:

```c
int main (void) {

  HAL_Init ( );
  SystemClock_Config ( );


  LED_Initialize ();              /* LED Initialization */
  ADC_Initialize_and_Set_IRQ ();/* ADC Special Init */

  while (1) {                     /* output 8-bit adcValue */
    LED_SetOut (adcValue >> 4);   /* to LEDs            */
    wait_delay ( 100 );                    /* wait */
  }
}
```

6. Create a `Custom_ADC.c` file and enter code to set up the ADC:

```c
#include ""stm32f4xx_hal.h"" /* STM32F4xx Definitions */
#include ""Custom_ADC.h""


/*-------------------------------------------------
 * ADC_Initialize_and_Set_IRQ: Initialize Analog to
 *              Digital Converter and Enable IRQ
 *-------------------------------------------------*/
void ADC_Initialize_and_Set_IRQ (void) {
    /* Setup potentiometer pin PF9 (ADC3_7) and ADC3 */

  RCC->APB2ENR |= (1UL <<  10);     /* En. ADC3 clk */
  RCC->AHB1ENR |= (1UL <<   5);     /* En. GPIOF clk */
  GPIOF->MODER |= (3UL << 2*9);/* PF9 is Analog mde */

  ADC3->SQR1   =    0;
  ADC3->SQR2   =    0;
  ADC3->SQR3   =   (7UL <<   0);    /* SQ1 = channel 7 */
  ADC3->SMPR1  =    0;              /* Channel 7 smple */
  ADC3->SMPR2  =   (7UL <<  18); /* time = 480 cyc. */
  ADC3->CR1    =   (1UL <<   8);     /* Scan mode on */
  ADC3->CR2    &= ~2;              /* single conv. mode */

  ADC3->CR1   |=  ( 1UL <<   5);      /* En. EOC IRQ */
  ADC3->CR2   |=  ( 1UL <<   0);       /* ADC enable */
  NVIC_EnableIRQ( ADC_IRQn );             /* En. IRQ */
  ADC3->CR2 |= (1 << 30);    /* Start 1st conversion */
}
```

7. Add the `adcISR.c` and `Custom_ADC.c` files to the project.

8. Declare a function prototype for `ADC_Initialize_and_Set_IRQ ()` in the `Custom_ADC.h` file.

9. Build, download, and run the program.

How it works...

The STM32F407xx features *3 × 12*-bit successive approximation ADCs, each sharing up to 16 external channels and performing conversions in single-shot or scan mode. A simplified schematic showing the architecture of each converter is presented next (please note that a more detailed diagram is included in STM's *RM0090* Reference manual at `http://www.st.com`).

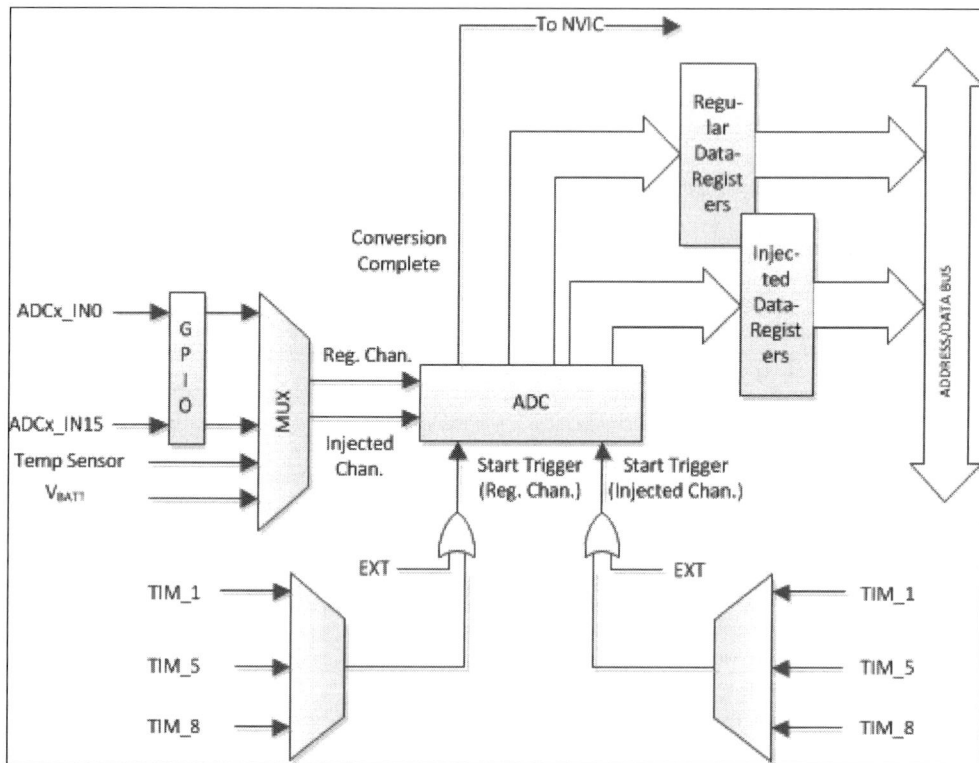

Simplified STM32F4xxxx microcontroller ADC schematic

The 16 multiplexed input channels are organized in two groups comprising regular and injected channels. A subset of GPIO port pins can be connected to the ADC multiplexer by configuring the pin as a high-impedance analog input. The pin/input channel mapping is device-dependent. Details for the *STM32F407IG* device used by the ARM MCBSTM32F400 evaluation board can be found in the STM32F405xx and STM32F407xx Datasheet (`http://www.st.com`), and a simplified form is given in the following table. The ADC can be configured to carry out a sequence of up to 16 conversions on each group, each triggered separately by either an external-or-timed start signal.

ADC1 Input Channel	GPIO Port	ADC2 Input Channel	GPIO Port	ADC3 Input Channel	GPIO Port
IN_0	PA0	IN_0	PA0	IN_0	PA0
IN_1	PA1	IN_1	PA1	IN_1	PA1
IN_2	PA2	IN_2	PA2	IN_2	PA2
IN_3	PA3	IN_3	PA3	IN_3	PA3
IN_4	PA4	IN_4	PA4	IN_4	PF6
IN_5	PA5	IN_5	PA5	IN_5	PF7
IN_6	PA6	IN_6	PA6	IN_6	PF8
IN_7	PA7	IN_7	PA7	IN_7	PF9
IN_8	PB0	IN_8	PB0	IN_8	PF10
IN_9	PB1	IN_9	PB1	IN_9	PF3
IN_10	PC0	IN_10	PC0	IN_10	PC0
IN_11	PC1	IN_11	PC1	IN_11	PC1
IN_12	PC2	IN_12	PC2	IN_12	PC2
IN_13	PC3	IN_13	PC3	IN_13	PC3
IN_14	PC4	IN_14	PC4	IN_14	PF4
IN_15	PC5	IN_15	PC5	IN_15	PF5

The GPIO ports used by the ADC must be configured as analog inputs by writing appropriate values to MODERy[1:0] bits of the Mode Register that is shown as follows:

31	30	29	28	27	26	25	24
MODER15[1:0]		MODER14[1:0]		MODER13[1:0]		MODER12[1:0]	
rw	rw	rw	rw	rw	rw	rw	rw
23	22	21	20	19	18	17	16
MODER11[1:0]		MODER10[1:0]		MODER09[1:0]		MODER08[1:0]	
rw	rw	rw	rw	rw	rw	rw	rw
15	14	13	12	11	10	09	08
MODER07[1:0]		MODER06[1:0]		MODER05[1:0]		MODER04[1:0]	

rw	rw	rw	rw	rw	rw	rw	rw
07	06	05	04	03	02	01	00
MODER03[1:0]		MODE02[1:0]		MODER01[1:0]		MODER00[1:0]	
rw	rw	rw	rw	rw	rw	rw	rw

The Mode Register bits are defined as follows:

MODERy[1:0]	I/O Mode
00 :	Input
01 :	General Purpose output
10 :	Alternate Function
11 :	Analog Input

The ADC is configured by an initialization function named `ADC_Initialize_and_Set_IRQ()` that has been written specially for this recipe. The following description should be read with reference to STM"s *RM0090* Reference manual (`http://www.st.com`).

The thumbwheel labelled **ADC1** on the evaluation board provides a variable voltage input connected to GPIO port F pin 9 (ADC3 channel 7). To sample this voltage, we first configure `GPIOF` pin 9 as an analog input by writing to the port mode register (`GPIOF_MODER`). Statements in `ADC_Initialize_and_Set_IRQ()` are explained as follows:

1. The bit map for the port mode register shown in the MODER register bit table indicates that we must write logic-1 to bit 18 and 19. ARM writes the code like this to emphasize that we're configuring port F bit-9 (PF9):

   ```
   GPIOF->MODER |= (3UL << 2*9);
   ```

2. We also need to select the clock for `ADC3` and `GPIOF`:

   ```
   RCC->APB2ENR |= (1UL << 10);
   RCC->AHB1ENR |= (1UL <<  5);
   ```

3. Our aim is to set up a single conversion in the regular sequence. The first conversion is identified by the bits 4:0 of ADC regular sequence register 3 (`ADC_SQR3`). As PF9 maps to ADC3 channel 7, we write 7 to this register and 0 to `ADC_SQR1` and `ADC_SQR2`:

   ```
   ADC3->SQR1   =   0;
   ADC3->SQR2   =   0;
   ADC3->SQR3   =   (7UL <<  0);
   ```

4. The Sample/Hold time can be set (for each channel) by writing to the two ADC Sample Time Registers (`SMPR1` and `SMPR2`). In this case, as the input voltage is derived from a potentiometer, the sample frequency can be quite low, and so, a long Sample/Hold time of 480 cycles can be set:

```
ADC3->SMPR1   =    0;
ADC3->SMPR2   =   (7UL <<  18);
```

5. We carry out a single conversion on each group of channels identified by the regular sequence register, so we enable scan mode by writing to bit-8 of Control Register 1:

```
ADC3->CR1   =   (1UL <<  8);
```

6. To set up single conversion mode, enable an **end of conversion interrupt** (**EOCIE**), and enable the ADC (ADON), we write the following code:

```
ADC3->CR2   &=   ~2;
ADC3->CR1   |=   ( 1UL <<  5);
ADC3->CR2   |=   ( 1UL <<  0);
```

7. Finally, we must configure the Nested Vectored Interrupt Controller (NVIC) to respond to interrupts from the ADC and initiate the first conversion by writing `SWSTART` (bit-30), as follows:

```
NVIC_EnableIRQ( ADC_IRQn );
ADC3->CR2 |= (1 << 30);
```

The `ADC_IRQHandler` () interrupt handler needs to clear the interrupt, read the ADC data, and start another conversion cycle. The super-loop in the main function calls the `LED_SetOut` () function to display the most significant 8-bits of the ADC output on the LEDs.

There's more...

In continuous conversion mode, the ADC starts a new conversion as soon as the previous one has been completed. In practice, the new conversion starts after a delay of 15 cycles to allow the ADC to stabilize. Only the regular group of channels can be converted in continuous mode, as follows:

1. We can enable continuous mode by changing the last line of the function, `ADC_Initialize` (), to the following:

```
ADC3->CR2   |=   2;
```

2. As our interrupt handler no longer needs to trigger a new conversion, we only need the following two statements:

```
ADC3->SR &= ~2;
adcValue = (ADC3->DR);
```

Configuring general-purpose timers

The idea of this recipe, which we'll call `timerISR_c5v0`, is to use a general purpose timer (TIM2) to generate an interrupt every 100 ms (that is, 10 Hz). The interrupt handler maintains a counter that, in turn, sets the global variables, `LEDOn`, `LEDOff`, which are used within `main ()` to flash the LEDs.

How to do it...

Follow the steps to configure general purpose timers:

1. Create a new recipe (folder) named `timerISR_c5v0`. Invoke uVision5 and create a new project named `timerISR.uvprojx`.

2. Select the **LED (API)** driver from the RTE Board Support drop-down menu and configure **CMSIS** and **Device** options as in previous projects.

3. Create a new file, name it `timerISR.c`, and enter the following statements. Remember to include the boilerplate:

```
#include ""stm32f4xx_hal.h""
#include ""Board_LED.h""
#include <stdbool.h>
#include ""timer.h""

/* Globals */
uint32_t tic = 0;

#ifdef __RTX


/* Function Prototypes */
void SystemClock_Config(void);

/**
  * System Clock Configuration
  */
void SystemClock_Config(void) {

```

4. Define a handler for the timer interrupt by adding these statements to the `timerISR.c` file:

```
void TIM2_IRQHandler (void) {

    /* check IRQ source */
    if ((TIM2->SR & 0x0001) != 0) {
        tic++;
```

```
        TIM2->SR &= ~(1<<0);              /* clear UIF flag */
    }
}
```

5. Define a `main ()` function in the `timerISR.c` file:

```c
int main (void) {
    int32_t num = 0;
    uint32_t toc;
    uint32_t count = 0;

    HAL_Init ( );
    SystemClock_Config ( );

    TIM2_Initialize ( );/* Gen. interrupt each 100 ms */
    LED_Initialize();              /* LED Initialization */

    while (1) {
        if (toc != tic) {
            toc = tic;
            LED_Off (num);
        if (count < 7)
            num = (num+1);
        else
            num = (num-1);
        LED_On (num);
        count = (count+1)%14;
        }
    }
}
```

6. Open a new file, add the following source code, save the file, and name it `timer.c`:

```
/*-------------------------------------------------
 * Recipe:  timerISR_c5v0
 * Name:    timer.c
 * Purpose: Low level timer functions
 *-------------------------------------------------
 *
 * Modification History
 * 16.04.14 created
 * 22.12.15 updated (uVision5.17+DFP2.6.0)
 *
 * Mark Fisher, CMP, UEA, Norwich
 *-----------------------------------------------*/
```

```
#include ""stm32f4xx_hal.h""          /* STM32F4xx Defs */
#include ""timer.h""


/*********************************************************
 * TIM2_Initialize ( )
 *********************************************************
 * Initializes TIM2 generates interrupts every 100ms (0.1s)
 * SystemCoreClock = 168 MHz - set by SystemInit ( )
 * Refer to Figure 134 of STM Reference Manual RM0090
 * TIMxCLK = SystemCoreClock/2
 * Hence ticks = 0.1 * 168,000,000 / 2 = 8,400,000
 * Prescaler = 8400-1; ARR = 1000-1;
 *********************************************************/
void TIM2_Initialize (void) {
  const uint16_t PSC_val = 8400;
  const uint16_t ARR_val = 1000;

  RCC->APB1ENR |= RCC_APB1ENR_TIM2EN; /* En TIM2 clk */

  TIM2->PSC = PSC_val - 1;              /* set prescaler */
  TIM2->ARR = ARR_val - 1;             /* set auto-reload */
  TIM2->CR1 = (1UL << 0);            /* set command reg. */
  TIM2->DIER = (1UL << 0);             /* Enable TIM2 IRQ */
  NVIC_EnableIRQ(TIM2_IRQn);           /* En. NVIC TIM2 IRQ */
}
```

7. Add `timer.c` and `timerISR.c` to the project.

8. Create a suitable header file named `timer.h` containing function prototypes for `timer.c`.

9. Build, download, and run the program.

How it works...

As microcontrollers were conceived to target real-time applications, counter-timers have always been a prominent feature of their architecture. Timers can be used for a variety of purposes, including measuring pulse lengths of input signals, generating output signals, triggering interrupts, or other events. The STM32F407xx microcontroller family that is used by the evaluation board provides 14 timers (TIM1-TIM14).

Type	Size	Identifier
Advanced Control Timers	16-bit	TIM1, TIM8
General Purpose Timers	16/32-bit	TIM2-TIM5

Type	Size	Identifier
Basic Timers	16-bit	TIM6, TIM7
General Purpose Timers	16-bit	TIM9-TIM14

A simplified schematic for general purpose timers is shown in the following diagram (a more detailed schematic can be found in STM's *RM0090* Reference manual at http://www.st.com).

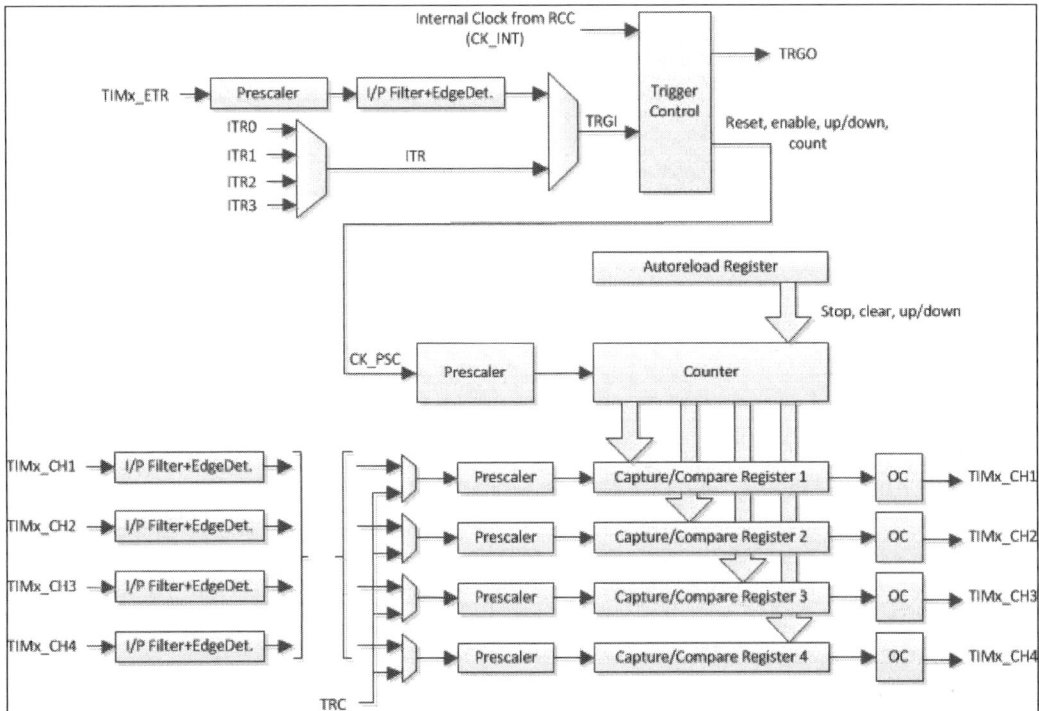

Advanced timers, **TIM1** and **TIM8**, provide similar functionality and include some additional features, such as a repetition counter, break inputs, and complementary outputs with programmable dead time. These are useful when implementing complex **pulse width modulation** (**PWM**) schemes. The main component is the time-base unit comprising a 16/32-bit counter and its related auto-reload register and prescaler. The prescaler clock (CK_PSC) can be selected from one of the following:

- ▶ **Internal clock** (CK_INT): This is derived from the reset and clock control (RCC) peripheral.
- ▶ **External clock mode 1:** This is the External input pin (TIx)
- ▶ **External clock mode 2:** The External trigger input (ETR) is available on TIM2, TIM3, and TIM4, only
- ▶ **Internal trigger inputs (ITRx):** This allows one timer to act as a prescaler for another

Following RESET, the CK_INT internal clock is selected. CK_INT is derived from the APBx timer output of the **Reset and Clock Control** (**RCC**) unit; refer to STM's *RM0090* Reference manual, Figure 21, (`http://www.st.com`). The timer clock frequencies are set automatically by hardware. The frequency depends on the setup used for the APB domain prescaler. There are two cases, as follows:

- If the APB prescaler is 1, the timer clock frequencies are set to the same frequency as that of the APB domain to which the timers are connected

- Otherwise, they are set to twice (×2) the frequency of the APB domain to which the timers are connected

The RCC unit manages all the clocks used by the microcontroller. The **system clock (SYSCLK)** can be derived from one of three sources:

- HSI clock

- HSE clock

- PLL clock

The `SystemInit( )` function defined in the `system_stm32f4xx.c` file is called by the `startup_stm32f4xx.s` file to configure the system clock before branching to the main program. The `SystemCoreClock` global variable is assigned a value representing the SYSCLK frequency and is available to user applications (for example, to set the SysTick timer). `SystemInit()` also configures the AHB and APB domain prescalers.

The internal (HIS) clock and external crystal-controlled oscillator (HSE) clock are connected to the main phase locked loop (PLL) within the microcontroller and this provides two outputs:

- The first output is used to generate the high-speed system clock (upto 168 MHz)

- The second output is used to generate the clock for the USB OTG FS (48 MHz), the random analog generator (≤48 MHz), and the SDIO (≤48 MHz)

The MCBSTM32F400 evaluation board uses a 25 MHz external oscillator, which gives a PLL frequency of 168 MHz, and `SystemInit ()` selects this as SYSCLK.

The main component of the time-base unit is a 16-bit or 32-bit counter (CNT) and its associated auto-reload register (ARR). The counter clock can be divided by a prescaler (PSC). Both the counter, prescaler, and auto-reload register can be written or read by software. The prescaler can divide the counter clock frequency by any factor between 1 and 65,536 (2^{16}). The operation of the counter and auto-reload register depends on the how the counter is configured. Three configuration modes are available, named upcounter, downcounter, and center-aligned. The timing diagram shown next illustrates the upcounter mode with the prescaler set to divide by 2 (other modes are described in the *RM0090* Reference manual, `http://www.st.com`). In upcounting mode, the counter counts from 0 to the auto-reload value (the content of the TIMx_ARR register), then restarts from 0 and generates a counter overflow event.

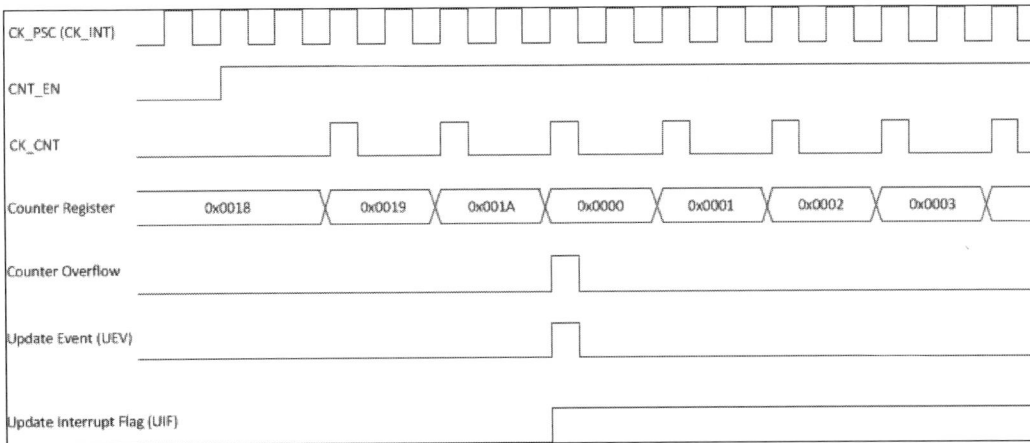

The steps required to configure TIM2 are as follows:

1. First enable the `TIM2` clock by writing to the `RCC` APB1 Enable Register:

   ```
   RCC->APB1ENR |= RCC_APB1ENR_TIM2EN;
   ```

   The number of SYSCLK ticks in 0.1 s can be found by:

   $$0.1 \times \frac{\text{SYSCLK}}{2}$$

   when SYSCLK = 168 MHz this gives a value of 8,400,000, which is achieved by a prescaler value of 8,400 and auto-reload register value of 1,000, that is, as follows:

   ```
   const uint16_t PSC_val = 8400;
   const uint16_t ARR_val = 1000;
   ```

   The prescaler divides the input clock by a factor PSC[15:0] +1:

   $$CK\_CNT = f_{CK\_PSC}/PSC[15:0] + 1$$

   So we write the following

   ```
   TIM2->PSC = PSC_val - 1;
   ```

2. Similarly, as the counter is reset to zero, we write the following:

   ```
   TIM2->ARR = ARR_val -1;
   ```

3. Then, enable the counter and enable interrupts:

   ```
   TIM2->CR1 = (1UL << 0);
   TIM2->DIER = (1UL << 0);
   ```

4. Finally, configure the Nested Vectored Interrupt Controller to respond to TIM2 interrupts:

```
NVIC_EnableIRQ(TIM2_IRQn);
```

Once configured, Timer 2 generates interrupts every 100 ms, and the interrupt handler increments a counter (tic). The code within the superloop generates a visually interesting pattern.

Using timers to trigger conversions

As sampling frequency plays such a critical role in determining the quality of the digital representation of the analog signal input, and to avoid aliasing artifacts, it is preferable to use a timer to trigger the conversion rather than to enable continuous conversions as we did in the previous recipe. This recipe, adcTimerISR_c5v0, illustrates this technique. The aim of this recipe is to configure TIM2 _CH2 in output compare mode so that it toggles every 100 ms and then use this timing signal to trigger the ADC.

How to do it...

1. First create a new project called adcTimerISR.uvprojx and use the RTE manager to configure it as we did for the folder adcISR_c5v0 for the *Setting up the ADC* recipe.

2. Copy timer.c and Custom_ADC.c from the previous recipes and add these to the project. Copy adcISR.c and rename it adcTimerISR.c. Add this to the project.

3. Add #include timer.h to adcTimerISR.c and call TIM2_Initialize() in main(). Check whether the project successfully builds.

4. Modify the TIM2_Initialize() function so that it no longer produces an update interrupt flag (UIF) by deleting the following statements:

```
TIM2->DIER = (1UL << 0);
NVIC_EnableIRQ(TIM2_IRQn);
```

5. Configure TIM2_CH2 to toggle channel 2 capture/compare output by writing to the appropriate fields of Capture/Compare Mode Register 1 (CCMR1) and Capture/Compare Enable Register (CCER):

```
TIM2->CCMR1 |= ( 3UL << 12 );
TIM2->CCER  |= ( 1UL << 4 );
```

> There is no need to write to the Capture/Compare Register. If we leave it set to zero (that is, Reset), then the Capture/Compare output will toggle each time TIM2_CNT is zero (that is, every 100 ms):

```
/******************************************************
 * TIM2_Initialize ( )
 ******************************************************
 * Initializes TIM2
 * Capture Compare 2 Interrupt Flag (CC2IF)
 * generates interrupts every 100ms (0.1s)
 * SystemCoreClock = 168 MHz - set by SystemInit ( )
 * Refer to Figure 134 of STM Reference Manual
 * TIMxCLK = SystemCoreClock/2
 * Hence ticks = 0.1 * 168,000,000 / 2 = 8,400,000
 * Prescaler = 8400-1; ARR = 1000-1;
 * Note: Capture Compare Register is left in Reset
 ******************************************************/
void TIM2_Initialize (void) {
  const uint16_t PSC_val = 8400;
  const uint16_t ARR_val = 1000;

  /* En. clk for TIM2 */
  RCC->APB1ENR |= RCC_APB1ENR_TIM2EN;

  TIM2->PSC = PSC_val - 1;        /* set prescaler */
  TIM2->ARR = ARR_val - 1;       /* set auto-reload */
  TIM2->CR1 = ( 1UL << 0 );        /* set Ctr. En. (CEN) */
  TIM2->CCMR1 |= ( 3UL << 12 );   /* OC1REF toggles
                                               when TIMx_
CNT=TIMx_CCR1*/
  TIM2->CCER |= ( 1UL << 4 );      /* CC2E set */
}
```

6. Modify the `adc_Initialize_and_Set_IRQ ( )` function to trigger conversions on both the rising and falling edge of `TIM2_CH2` by writing to Control Register 2:

```
    ADC3->CR2    |=  ( 3UL << 28);
    ADC3->CR2    |=  ( 3UL << 24);
```

7. Remember to run the ADC in single conversion mode:

```
void ADC_Initialize_and_Set_IRQ (void) {
    /* Setup potentiometer pin PF9 (ADC3_7) and ADC3 */

  RCC->APB2ENR |= (1UL << 10);       /* En. ADC3 clk */
  RCC->AHB1ENR |= (1UL << 5);       /* En. GPIOF clk */
  GPIOF->MODER |= (3UL << 2*9);/* PF9 = Analog mode */

  ADC3->SQR1   =   0;
  ADC3->SQR2   =   0;
```

```
ADC3->SQR3    =   (7UL <<  0);        /* SQ1 = chan. 7 */
ADC3->SMPR1   =   0;                  /* Chan. 7 sample */
ADC3->SMPR2   =   (7UL <<  18); /* time = 480 cyc. */
ADC3->CR1     =   (1UL <<  8);        /* Scan mode on */

ADC3->CR1    |=  ( 1UL <<  5);        /* En. EOC IRQ */
ADC3->CR2    |=  ( 3UL << 28); /* Trig on both edg */
ADC3->CR2    |=  ( 3UL << 24);        /* of TIM2_CC2 */
ADC3->CR2    |=  ( 1UL <<  0);        /* ADC enable */
NVIC_EnableIRQ( ADC_IRQn );           /* Enable IRQ */
ADC3->CR2 |= (1 << 30);   /* Start 1st conversion */
}
```

8. Build, download, and run the program. You will notice that, when we execute this program, the output appears much more stable than it did using a continuous mode. This is just a consequence of performing fewer conversions, but it does serve to emphasize the need to avoid oversampling unless there is good reason.

How it works...

In addition to the update event interrupt, each timer also allows interrupts to be generated by up to four capture compare channels (TIMx_CH1-TIMx_CH4). Each Capture/Compare channel comprises a Capture/Compare register, an input stage for capture (with digital filter, multiplexing, and prescaler), and an output stage (with comparator and output control). Each can be configured as the input capture, PWM input, forced output, output compare, PWM, or one-pulse modes. The output compare mode can be used to provide timing signals that can be used to start A-D conversions.

One of 16 possible start conversion triggers can be selected for the regular group of channels by writing to the ADC control register 2 (ADC_CR2) bit field, EXTSEL[3:0]. The following table shows how the trigger sources are encoded

> *CH1-CH4* and *TRGO* refer to timer channels. For further information, refer to STM's *RM0090* Reference manual (http://www.st.com), Chapters *17* and *18*.

EXTSEL[3:0]	Start Trigger	EXTSEL[3:0]	Start Trigger
0000	TIM1_CH1	1000	TIM3_TRGO
0001	TIM1_CH2	1001	TIM4_CH4
0010	TIM1_CH3	1010	TIM5_CH1
0011	TIM2_CH2	1011	TIM5_CH2

EXTSEL[3:0]	Start Trigger	EXTSEL[3:0]	Start Trigger
0100	TIM2_CH3	1100	TIM5_CH3
0101	TIM2_CH4	1101	TIM8_CH1
0110	TIM2_TRGO	1110	TIM8_TRGO
0111	TIM3_CH1	1111	EXT11

The *polarity* of the trigger is determined by *EXTEN*, as shown in the following table:

EXTEN	Trigger Polarity
00	Trigger detection disabled
01	Trigger detection on the rising edge
10	Trigger detection on the falling edge
11	Trigger detection on both the rising and falling edges

There's more...

If we wish to confirm that the ADC is sampled every 100 ms, then simply add a global tick variable and increment this in the IRQ handler. Change the code within the super-loop to blink the LEDs every 10 ticks.

Setting up the DAC

The aim of this recipe is to echo the analog voltage input to the ADC to the DAC. The DAC operation is relatively simple as compared to the ADC. The MCBSTM400 evaluation board doesn't provide any means of directly monitoring either of the DAC channels. As DAC channel 2 (output to PA5) drives the clock for the USB 2.0 transceiver (IC6), the only option that we have is to use DAC channel 1 (output PA4). To see an output, we'll need to probe the output PA4 with a test meter. This recipe is called `echo_adc_dac_c5v0`.

How to do it...

To set up the DAC follow the steps outlined:

1. Clone `adcTimerISR_c5v0` from the *Using timers to trigger conversions* recipe and extend it by adding the `dac.c` and `dac.h` files. These will be used to define a function called `DAC_Initialize()` (shown next) that will be used to set up the DAC; the DAC registers and mask definitions are defined as a data structure in the `stm32f4xx_hal.h` file:

```
#include ""stm32f4xx_hal.h""        /* STM32F4xx Defs */
```

```
#include ""DAC.h""

/*----------------------------------------------------------
 *        DAC_Initialize: Initialize DAC
 *
 * Parameters:   (none)
 * Return:       (none)
 *---------------------------------------------------------*/
void DAC_Initialize (void) {

    RCC->APB1ENR |= RCC_APB1ENR_DACEN; /* En. DAC clk */
                                       /* En. GPIOA clk */
    RCC->AHB1ENR |= RCC_AHB1ENR_GPIOAEN;
    GPIOA->MODER |= (3UL << 2*4);/* PA4 = Analog mode */

    DAC->CR |= DAC_CR_EN1;             /* Enable DAC 1 */
    DAC->CR |= DAC_CR_BOFF1;   /* Enable DAC 1 OP Buff */
}
```

2. Add dac.c to the project.

3. Add a function prototype to dac.h.

4. Modify the main() function to call the DAC_Initialize() function and add a statement in the main loop to write the ADC value to the DAC:

```
int main (void) {

    HAL_Init();
    SystemClock_Config();

    LED_Initialize ();                  /* LED Init.         */
    ADC_Initialize_and_Set_IRQ ();/* ADC Special Init. */
    DAC_Initialize ();                          /* DAC Init. */
    TIM2_Initialize ();                         /* TIM2 Init. */

    while (1) {                 /* output 8-bit adcValue */
        DAC->DHR12R1 = adcValue;        /* Echo ADC to DAC */
        LED_SetOut (DAC->DOR1 >> 4); /* Echo DOR to LEDs */
        }
}
```

5. Build, download, and run the program.

How it works...

The STM32F407xx features *2 x 12*-bit buffered DAC converter channels, DAC1 and DAC2. Eight DAC trigger inputs are provided for each device. The STM32F405xx and STM32F407xx Datasheet *Table 7* (`http://www.st.com`) shows that the DAC1 and 2 outputs are featured as an additional function of GPIO PA4 and PA5, respectively. The GPIO I/O port bit must be configured as analog to disable the GPIO output buffer. A simplified block diagram of a DAC channel is shown as follows (a more detailed diagram can be found in STM's *RM0090* Reference manual at `http://www.st.com`):

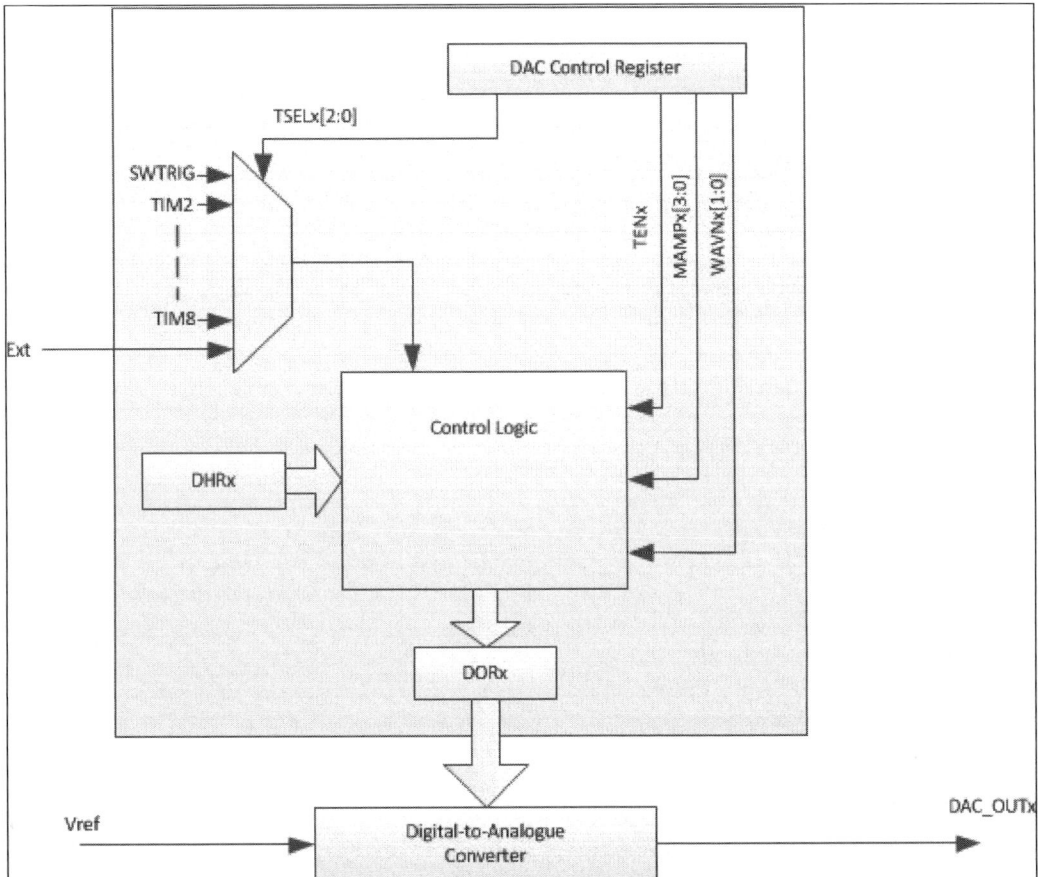

The DAC can be configured in 8- or 12-bit mode. In 12-bit mode, the data can be left- or right-aligned by writing to the appropriate **Data Holding Register** (**DHR**). The DAC **Data Output Register** (**DOR**) cannot be written to directly. Data is transferred from the DHR to the DOR after one APB1 clock if no trigger is selected; or if a trigger is selected, then the transfer occurs three APB1 clocks after the trigger event.

The `DAC_Initialize()` function performs the following operations:

1. The first step is to write to the Reset and Clock Control (RCC) peripheral and enable clocks for the DAC and GPIO port A.

2. To enable the DAC clock, we write bit-29 of the `APB1` peripheral clock enable register:

    ```
    RCC->APB1ENR |= RCC_APB1ENR_DACEN;
    ```

3. To enable Port A, clocks write bit-0 of the `AHB1` peripheral clock enable register:

    ```
    RCC->AHB1ENR |= RCC_AHB1ENR_GPIOAEN;
    ```

4. Then, we configure Port A bit-4 (PA4) in analog mode to source the analog output by writing to the mode register:

    ```
    GPIOA->MODER |= (3UL << 2*4);
    ```

5. Finally, we enable the `DAC` channel 1 and its associated output buffer. This step involves writing to the `DAC` control register:

    ```
    DAC->CR |= DAC_CR_EN1;
    DAC->CR |= DAC_CR_BOFF1;
    ```

We write a statement in the main loop to write the ADC value to the DAC. We use the simplest conversion mode that triggers a conversion each time data is written to the (DHR). There are three Data Holding Registers for each channel. Each loads the (DOR) slightly differently. We choose the DHR that loads the DOR with a right-aligned 12-bit value. Writing to the DHR is achieved by the following:

```
DAC->DHR12R1 = adcValue;
```

Instead of writing `adcValue` to the LEDs, we read the DAC DOR and write its value instead. Please note that the DOR is read-only (it cannot be written by software). Writing the LEDs in this way will confirm that we've correctly configured the DAC. If the DOR shows the correct value but there is no output voltage on PA4, then the problem lies with the GPIO Port configuration. The following statement writes the `DAC1` DOR value to the LEDs:

```
LED_Out (DAC->DOR1 >> 4);
```

There's more...

The DAC converter includes a **linear-feedback shift register** (**LFSR**) and can be configured to generate pseudo-random noise and a programmable triangle-wave generator is also available; refer to STM's *RM0090* Reference manual and STM, Application Note AN3216: Audio and waveform generation using the DAC in STM32 microcontroller families (`http://www.st.com`) for more details.

Generating a sine wave

Sinusoidal signals are commonly used in signal processing applications and generating these waveforms provides an interesting project that is the focus of this recipe. A common approach is a direct method that stores the sinusoidal waveform samples in a **look-up-table (LUT)**. This recipe is called `dacSinusoid_c5v0`.

Getting ready

First, we need to calculate the (12-bit) DAC values that will be stored in the LUT. We'll attempt to generate a 50 Hz sinusoidal signal and use a spreadsheet (for example, Microsoft Excel) to calculate the following values:

Smpl. No	Theta Rads	floor((sin(theta)+1)*4095/2)
0	0	2047
1	0.31415927	2680
2	0.62831853	3250
3	0.9424778	3703
4	1.25663706	3994
5	1.57079633	4095
6	1.88495559	3994
7	2.19911486	3703
8	2.51327412	3250
9	2.82743339	2680
10	3.14159265	2047
11	3.45575192	1414
12	3.76991118	844
13	4.08407045	391
14	4.39822972	100
15	4.71238898	0
16	5.02654825	100
17	5.34070751	391
18	5.65486678	844
19	5.96902604	1414

How to do it...

Follow the outlined steps to generate a sine wave:

1. Create a new recipe called `dacSinusoid_c5v0` by cloning `timerISR_c5v0` from the *Using timers to trigger conversions* recipe.

2. Replace `timerISR.c` with a file named `dacSinusoid.c` and add a declaration for an LUT:

    ```c
    uint16_t dacLUT [] = {2047, 2680, 3250, 3703, 3994,
                          4095, 3994, 3703, 3250, 2680,
                          2047, 1414,  844,  391,  100,
                          0,     100,  391,  844, 1414 };
    ```

3. Add an interrupt handler to service `TIM2`:

    ```c
    /*-----------------------------------------------
       TIM2 IRQ Handler
     *---------------------------------------------------*/
    void TIM2_IRQHandler (void) {
      static uint8_t idx = 0;

      if (TIM2->SR & (1<<0)) {
        TIM2->SR &= ~(1<<0);          /* clear UIR flag */
                                  /* write LUT val to DAC */
        DAC->DHR12R1 = dacLUT[idx++];
         idx %= 20;
         LED_Out (idx);           /* Write idx to LEDs */
        }
    }
    ```

    Add the following `main()` function:

    ```c
    /*-----------------------------------------------
       Main function
     *--------------------------------------------------*/
    int main (void) {

      HAL_Init();
      SystemClock_Config();

      LED_Initialize ();                    /* LED Init. */
      DAC_Initialize ();                     /* DAC Init */
      TIM2_Initialize ();

      while (1) {
         /* empty statement */  ;
        }
    }
    ```

4. Add `dacSinusoid.c` to the project.

5. Only one statement in the `TIM2_Initialize ( )` function (in the `timer.c` file) needs to be changed:

```
/*******************************************************
 * TIM2_Initialize ( )
 *******************************************************
 * Initializes TIM2 generates interrupts every 1ms
 * SystemCoreClock = 168 MHz - set by SystemInit ( )
 * Refer to Figure 134 of STM Reference Manual RM0090
 * TIMxCLK = SystemCoreClock/2
 * Hence ticks = 0.001 * 168,000,000 / 2 = 84,000
 * Prescaler = 84-1; ARR = 1000-1;
 *******************************************************/
void TIM2_Initialize (void) {
  const uint16_t PSC_val = 84;
  const uint16_t ARR_val = 1000;

                                  /* En. TIM2 clk */
  RCC->APB1ENR |= RCC_APB1ENR_TIM2EN;

  TIM2->PSC = PSC_val - 1;        /* set prescaler */
  TIM2->ARR = ARR_val - 1;        /* set auto-reload */
  TIM2->CR1 = (1UL << 0);         /* set command reg. */
  TIM2->DIER = (1UL << 0);        /* En. TIM2 IRQ */
  NVIC_EnableIRQ(TIM2_IRQn);      /* En. NVIC TIM2 Int. */
}
```

6. Build, download, and run the program.

How it works...

The many techniques that could be used to generate a sinusoidal waveform are the subject of the digital signal processing literature. A common approach is a direct method that stores the sinusoidal waveform samples in a look-up-table (LUT). This may seem very crude but if the output is passed through an analog low-pass filter with a cut-off frequency set to the fundamental frequency of the output signal, then the result is a reasonably pure sinusoid. In fact, this approach works equally well for a triangular waveform (which can be generated by the DAC hardware), but the LUT approach will produce something that looks convincing when displayed on an oscilloscope without the need for a filter.

In theory, the minimum number of samples needed is determined by the *Nyquist-Shannon Sampling Theorem*. This states that we need a minimum of two samples per cycle. At this limit the raw samples describe a 50 Hz square wave that will produce a sinusoid when processed by a suitable low-pass output filter. However, as an ideal square wave contains only components of odd-integer harmonic frequencies (of the form $2\pi(2k-1)f$), the order of the filter will need to be ~12 so that the harmonics are highly attenuated while the fundamental is unaffected. To achieve a satisfactory output with a much simpler second-order filter, the number of samples is usually increased by a factor of ~10.

We store the samples in an array, as follows:

```
uint32_t dacLUT [] = {2047, 2680, 3250, 3250, 3994,
                      4095, 3994, 3703, 3250, 2680,
                      2047, 1414,  844,  391,  100,
                         0,  100,  391,  844, 1414 };
```

Then, we use a timer to generate an interrupt every 1 ms (that is, the period of the sinusoid *T = 20 ms; 1/20 ms = 50 Hz.*). Please note that we could use any timer (in this case, we use `TIM2`; reusing code discussed previously but changing the prescaler value):

```
Uint16_t PSC_val = 84;
```

We write the sample to the DAC's Data Holding Register in the timer ISR (we postincrement `idx`), as follows:

```
DAC->DHR12R1 = dacLUT[idx++];
```

To ensure the index is incremented by modulo 20 (because the LUT array stores 20 values), we use the following:

```
idx %= 20;
```

We output the `idx` variable to the LEDs just to give a visual check that the program is running. A screenshot of an oscilloscope connected to PortA4 is shown as follows:

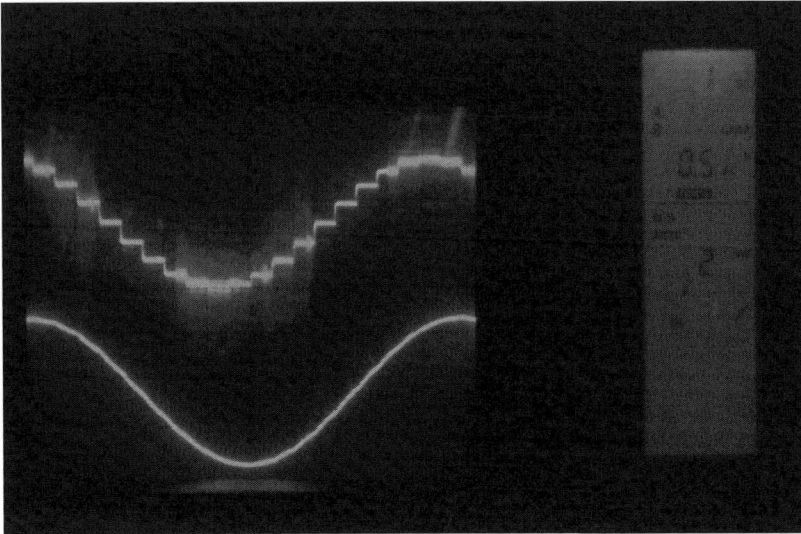

The lower trace shows the output (*Vout*) of the low-pass filter. The cut-off frequency for the low-pass filter is set to 50 Hz approximately, (refer to T. Floyd and D. Buchla, *Electronics Applications Circuits Devices and Applications (8e)*, Pearson Education, 2014) which can be seen in the following figure:

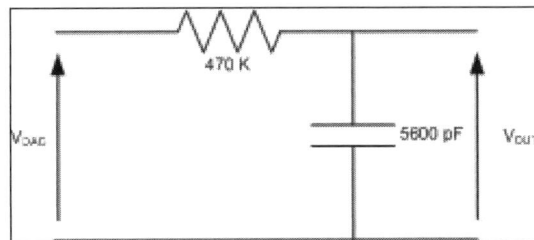

6

Multimedia Support

In this chapter, we will cover the following topics:

- ▶ Setting the RTE for the I2C Peripheral Bus
- ▶ How to use the LCD touchscreen
- ▶ Writing a driver for the audio codec
- ▶ How to use the audio codec
- ▶ How to use the camera
- ▶ Designing bitmapped graphics
- ▶ Ideas for games using sound and graphics

Introduction

Multimedia peripherals are discrete components that are connected to the microcontroller by a bus. Support for LCD touchscreens, audio codecs, and camera peripherals is a very attractive feature of the STM32F4xxx microcontroller, and selecting an evaluation board that includes these peripherals, although more expensive, will increase the range of projects that can be undertaken. Multimedia projects using the touchscreen and codec are great fun and much more likely to motivate young programmers than blinking LEDs. These peripherals are quite complex, but the libraries that are provided to support them are reasonably straightforward to use.

Setting the RTE for the I2C Peripheral Bus

The LCD touchscreen, three-axis motion sensor (LIS302DL), audio-codec (CS42L52), 64k EEPROM (M24C64), camera, and other peripherals that are supported by the MCBSTM32F400 evaluation board are connected to the STM32C microcontroller by a synchronous serial bus called I2C. The bus standard adopted is called the **Inter-Integrated Circuit (I2C)** Interface, which was developed by Phillips in the 1980s. Before we can use any peripherals that are connected to the I2C bus, we must first configure the I2C interface. We'll illustrate this by a recipe called `touchScreenDemo_c6v0`. Later in this chapter, we'll show you how to configure other I2C peripherals.

How to do it...

To set RTE for an I2C Peripheral Bus perform the following steps:

1. Open a new project (`touchScreenDemo`), in a new folder named `touchScreenDemo_c6v0`.

2. Using the RTE manager, select **Touchscreen** (an I2C peripheral) under **Software Component | Board Support**.

3. Set the **CMSIS** and **Device** options, as we've done for the previous recipes. Click **Resolve** and then **OK**:

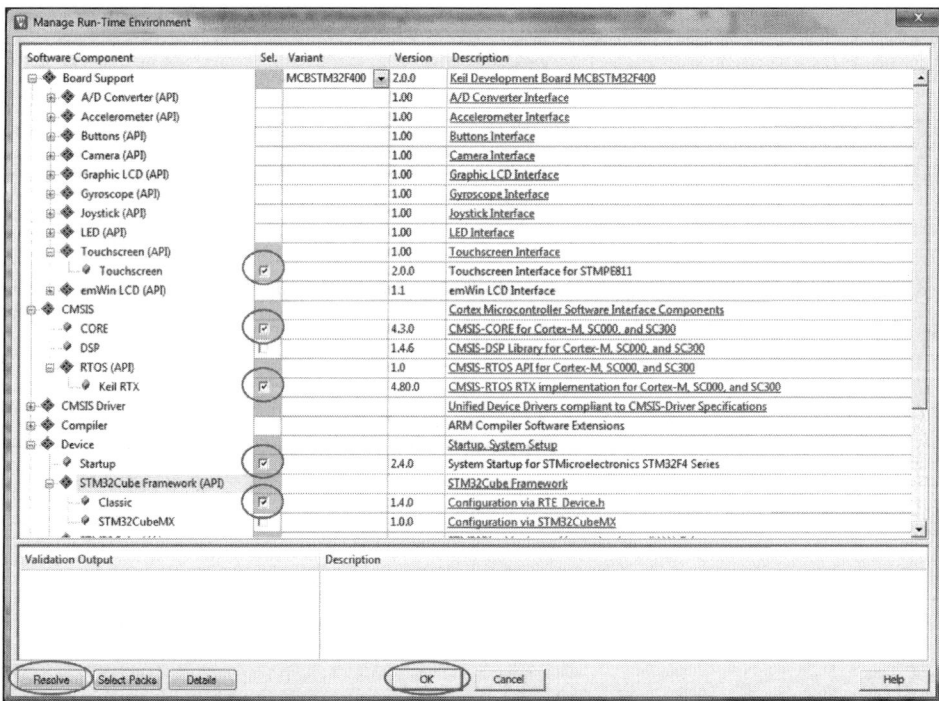

4. Open the `RTE_Device.h` file, select the **Configuration Wizard** editor tab, and enter the configuration choices that are shown in the following screenshot:

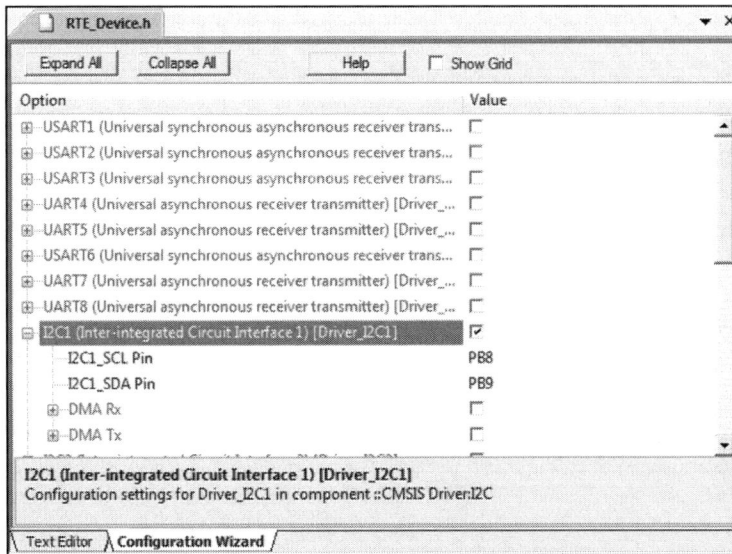

5. Open the `RTX_Conf_CM.c` file, select the **Configuration Wizard** editor tab, and enter the configuration choices that are shown in the following screenshot:

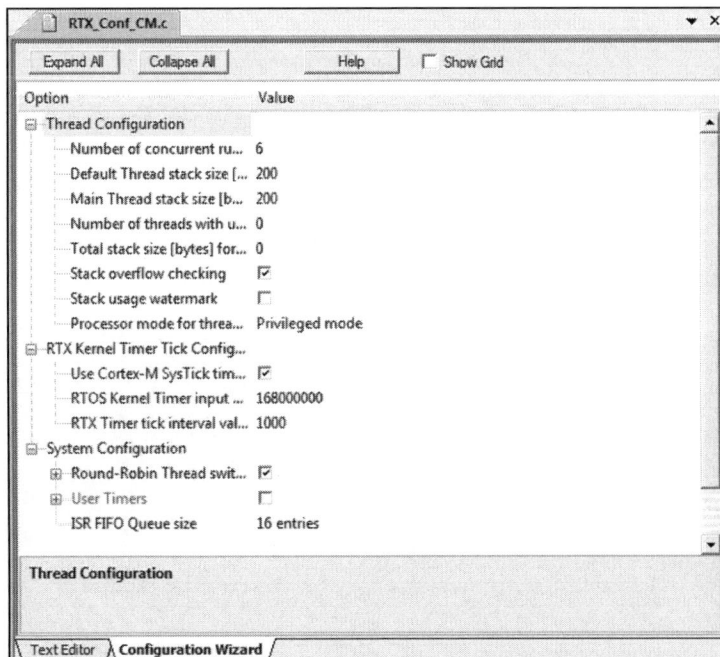

6. Check whether the program successfully compiles by declaring an empty main function (name the file, `touchScreenDemo.c`) and include this in the project:

```
int main (void) {
   HAL_Init ();        /* Init Hardware Abstraction Layer */
   SystemClock_Config ();              /* Config Clocks */
}
```

How it works...

A bus is the name that is given to a collection of signals (data, address, and control) that interconnect the processor infrastructure. The microcontroller uses a serial (rather than parallel) bus interconnection, and to keep the microcontroller pin count low, the bus signals are driven via a GPIO port that is configured in alternate function mode. I2C is a half-duplex synchronous serial bus comprising clock (SCL) and serial data (SDA) lines. Devices that are connected to the bus are identified by a 7- or 10-bit address and can be configured as master or slave. The following diagram shows a master node (in this case, the microcontroller) sourcing the clock and controlling slave devices connected to the bus (note that the master node does not have to be a microcontroller):

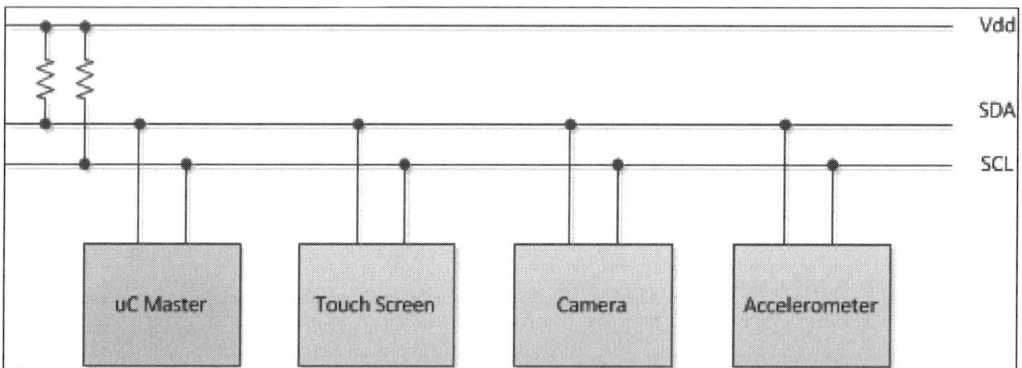

Before we can use the I2C bus, the bus master (that is, the microcontroller) must be configured. The MCBSTM32F400 evaluation board drives signals SDA and SCL via GPIO Port B bits 8 and 9, so before the interface can be used, GPIOB must be configured. This task is simplified using the uVision v5.x Run Time Environment (RTE) manager. To successfully compile a program that needs I2C, we must configure the `RTE_Device.h` file for our evaluation board. As we chose the **Device** option **STMCube_Framework → Classic**, the `RTE_Device.h` file for our evaluation board is provided by the RTE manager. A configuration wizard provides a simple user interface that allows different peripherals and parameters to be selected by tick boxes and drop-down lists. (Note that the Board Schematic confirms GPIO bits PB8 and PB9 are used to source signals, SDA and SCL.)

Accurate control of bus timing is critical for successful operation of the I2C. The RTE solves this by using a real-time kernel called RTX (we'll meet RTX in *Chapter 8, Real-Time Embedded Systems*). The Configuration Wizard for the `RTX_Conf_CM.c` file establishes certain scheduling parameters for the kernel.

Another serial interface standard supported by the MCU is known as **Serial Peripheral Interface** (**SPI**) and was developed by Motorola. For further information on I2C and SPI, refer to `http://www.byteparadigm.com/applications/introduction-to-i2c-and-spi-protocols/`.

How to use the LCD touchscreen

The LCD touchscreen used by the MCBSTM32F400 evaluation board is a resistive film giving a resolution of 4000 × 4000 (that is, far greater than the GLCD). This recipe extends `touchScreenDemo_c2v0` and illustrates how to use the LCD touchscreen.

How to do it...

Perform the following steps to use the LCD touchscreen:

1. Return to `touchScreenDemo_c2v0` and open the project.

2. Use the RTE manager to add **Software Component** → **Board Support** for the **Graphic LCD** (in addition to the Touchscreen). Click **Resolve** and then **OK**.

3. Open `touchScreenDemo.c`, and include the following headers:
    ```c
    #include <stdio.h>
    #include "stm32f4xx_hal.h"
    #include "cmsis_os.h"
    #include "Driver_I2C.h"
    #include "Board_GLCD.h"
    #include "GLCD_Config.h"
    #include "Board_Touch.h"
    ```

4. Define the following macros, global variables, and function prototypes:
    ```c
    // The size of the touch-screen co-ordinates system.
    #define SCREEN_TS_WIDTH  4000
    #define SCREEN_TS_HEIGHT 4000

    #define wait_delay HAL_Delay

    /* Globals */
    ```

```
        extern GLCD_FONT        GLCD_Font_16x24;


        /* Function Prototypes */
        void screenTransformTS(TOUCH_STATE *ts);

        void SystemClock_Config(void);

        void setDisplay(void);

        void updateDisplay(TOUCH_STATE  *tsc_state);

        void clearDisplay(void);
```

5. Extend the `main()` function:

```
    /*--------------------------------------------------
      Main function
     *-------------------------------------------------*/
    int main (void) {
      TOUCH_STATE  tsc_state;

      HAL_Init ();    /* Init Hardware Abstraction Layer */
      SystemClock_Config ();              /* Config Clocks */

      Touch_Initialize();  /* Touchscrn Controller Init */
      GLCD_Initialize();        /* Graphical Display Init */
      setDisplay();                /* Draw GLCD Display */

      while (1) {
        Touch_GetState (&tsc_state); /* Get touch state */

        if (tsc_state.pressed)
          updateDisplay(&tsc_state);
        else
          clearDisplay();

        wait_delay(100);
      }
    }
```

6. Add the `setDisplay()` function to `touchScreenDemo.c` file:

```
    /*--------------------------------------------------
      setDisplay
     *-------------------------------------------------*/
    void setDisplay( ) {
      GLCD_SetBackgroundColor (GLCD_COLOR_WHITE);
      GLCD_ClearScreen ();  /* clear the GLCD */
```

```
    GLCD_SetBackgroundColor (GLCD_COLOR_BLUE);
    GLCD_SetForegroundColor (GLCD_COLOR_WHITE);
    GLCD_SetFont (&GLCD_Font_16x24);
    GLCD_DrawString (0, 0*24, " CORTEX-M4 COOKBOOK ");
    GLCD_DrawString (0, 1*24, "  PACKT Publishing  ");

    GLCD_SetBackgroundColor (GLCD_COLOR_WHITE);
    GLCD_SetForegroundColor (GLCD_COLOR_BLACK);

    GLCD_DrawString (0, 3*24, "Touch:");
    GLCD_DrawString (0, 4*24, "x    :");
    GLCD_DrawString (0, 5*24, "y    :");
    GLCD_DrawString (0, 6*24, "xt   :");
    GLCD_DrawString (0, 7*24, "yt   :");
}
```

7. Add the `updateDisplay()` function to file `touchScreenDemo.c`:

```
/*-----------------------------------------------------
  updateDisplay
  *--------------------------------------------------*/
void updateDisplay(TOUCH_STATE  *tsc_state) {
  char buffer[128];

  GLCD_SetForegroundColor (GLCD_COLOR_BLACK);
  GLCD_DrawString (7*16, 3*24, "DETECTED");
  sprintf(buffer, "%i   ", tsc_state->x); /* raw x_coord */
  GLCD_DrawString (7*16, 4*24, buffer);

  sprintf(buffer, "%i   ", tsc_state->y); /* raw y_coord */
  GLCD_DrawString (7*16, 5*24, buffer);

  screenTransformTS(tsc_state);
  sprintf(buffer, "%i   ", tsc_state->x);
  GLCD_DrawString (7*16, 6*24, buffer);

  sprintf(buffer, "%i   ", tsc_state->y);
  GLCD_DrawString (7*16, 7*24, buffer);
}
```

8. Add the `clearDisplay()` function to file `touchScreenDemo.c`:

```
/*-----------------------------------------------------
  clearDisplay
  *--------------------------------------------------*/
void clearDisplay() {
  GLCD_SetForegroundColor (GLCD_COLOR_LIGHT_GREY);
  GLCD_DrawString (7*16, 3*24, "DETECTED");
```

```
        GLCD_DrawString (7*16, 4*24, "        ");
        GLCD_DrawString (7*16, 5*24, "        ");
        GLCD_DrawString (7*16, 6*24, "        ");
        GLCD_DrawString (7*16, 7*24, "        ");
}
```

9. Add the `screenTransformTS()` function to file `touchScreenDemo.c`:

```
/*-----------------------------------------------------
   Touch Screen Transform
 *---------------------------------------------------*/
void screenTransformTS(TOUCH_STATE *ts) {

    int y = ts->y;
    int x = ts->x;
    // Note: co-ordinates are inverted
    if (x > 0)
       ts->y = GLCD_HEIGHT - (int)(((double)x /
              (double)SCREEN_TS_HEIGHT)*(double)GLCD_HEIGHT);
    if (y > 0)
       ts->x = (int)(((double)y /
              (double)SCREEN_TS_WIDTH)*(double)GLCD_WIDTH);
}
```

10. Check the **Use MicroLIB** project option.

11. Build the project, download it, and run the program. The GLCD will display the LCD touchscreen and screen coordinates when touched (refer to the following screenshot):

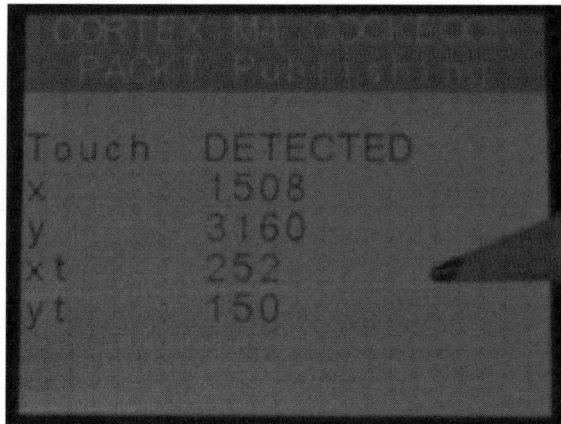

How it works...

The `Touch_GetState()` function updates the `tsc_state` variable, which stores the status of the LCD touchscreen and coordinates. These are stored as a structure that is defined by a `typedef` keyword in the `Board_Touch.h` file:

```
/* Touch state */
typedef struct _TOUCH_STATE {
  int16_t x;                          ///< Position X
  int16_t y;                          ///< Position Y
  uint8_t pressed;                    ///< Pressed flag
} TOUCH_STATE;
```

The LCD touchscreen and GLCD coordinate systems are different in resolution and origin. The `screenTransformTS( )` function maps between GLCD and touchscreen coordinate systems. Notice how we pass a pointer to the `tsc_state` variable and access specific fields such as `ts->y`, and so on.

Writing a driver for the audio codec

The audio codec is a peripheral that enables an analog signal to be converted and coded to a digital data stream or conversely the data stream to be decoded and converted back to an analog signal (`https://en.wikipedia.org/wiki/Codec`). The MCBSTM32F400 evaluation board uses a CS42L52 device that is manufactured by Cirrus Logic (`http://www.cirrus.com/en/products/cs42l52.html`). As, this codec is not yet included in Board Support, and as no CMSIS-compliant device driver is available, we are faced with the task of having to write our own driver.

However, this is not as daunting as it first appears because the code to set up and manage data transfer across the I2C bus can be lifted from the previous recipe (the `Touch_STMPE811.c` file) and the configuration of the CS42L52 codec is described in the data sheet. The recipe to develop and test this codec driver is called `codecDemo_c6v0`.

How to do it...

Perform the following steps to write a driver for the audio codec:

1. Create a new project called `codecDemo`, and using the Run-Time Environment manager, include **Board Support** for the **Graphic LCD**. Remember to configure Software Support for **CMSIS** and **Device** as in earlier projects.

2. Create a new file named `codecDemo.c`. Add the boilerplate to configure clocks, and so on, and a skeleton `main()` function:

```
int main (void) {
```

```
HAL_Init ();    /* Init Hardware Abstraction Layer */
SystemClock_Config ();              /* Config Clocks */
}
```

3. Add the `#include` files for the `codecDemo.c` file:

```
#include "stm32f4xx_hal.h"
#include "cmsis_os.h"
#include "codec_CS42L52.h"
#include "GLCD_Config.h"
#include "Board_GLCD.h"
#include <stdio.h>
```

4. Create a new file called `timer.c` and add this to the source code group. Add a function named `TIM3_Initialize()` to this file:

```
void TIM3_Initialize (void) {
  const uint16_t ARR_val = 7;

  /* enable clock for TIM3 */
  RCC->APB1ENR |= RCC_APB1ENR_TIM3EN;

  TIM3->CCMR1 = 0x00000070;        /* Set PWM Mode 2 */
  TIM3->ARR = ARR_val - 1;         /* set auto-reload */
  TIM3->CCR1 = 3;                  /* Duty cycle (~50%) */
  /* Enable capture/compare on Chan 1    */
  TIM3->CCER = 0x000B00001;
  TIM3->CR1 = 0x000B00001;             /* Enable counter */
}
```

5. Create a new file called `codec_CS42L52.c` and add this to the source code group. Copy the first 75 lines of the `Touch_STMPE811.c` file to `codec_CS42L52.c`, the first part of the file, including the `Touch_Read()` and `Touch_Write()` functions.

6. Change the `#include` directives in the `codec_CS42L52.c` file to the following:

```
#include "CS42L52.h"
#include "codec_CS42L52.h"
#include "stm32f4xx_hal.h"
#include "Driver_I2C.h"
#include "timer.h"
```

7. Replace any references to `TSC_I2C_ADDR` with `CODEC_I2C_ADDR`.

8. Replace any references to `TSC_I2C_PORT` with `CODEC_I2C_PORT`.

9. Replace `TSC_I2C_ADDR` with that given in the CS42L52 data sheet, as follows:

```
/* 7-bit I2C Address = 1001010b */
#define CODEC_I2C_ADDR      0x4A
```

10. Rename the `Touch_Read()` and `Touch_Write()` functions to `Codec_Read()` and `Codec_Write()`, respectively.

11. Add a global `typedef` to the `codec_CS42L52.c` file:

```
/* Global TypeDef - Register value */
typedef struct {
  uint8_t Addr;
  uint8_t Val;
} REG_VAL;
```

12. Add a function named `configureCodec()` to the `codec_CS42L52.c` file. The first two statements of configureCodec power the device down and wait for 10 ms. Note `#define delay_ms HAL_Delay`:

```
void configureCodec ( ) {
  Codec_Write(0x02, 0x01);    /* Keep Codec Power-down */
  delay_ms(10);

  for (i = 0; i < ARR_SZ(CODEC_Config_Init); i++)
    Codec_Write (CODEC_Config_Init[i].Addr,
      CODEC_Config_Init[i].Val);

  for (i = 0; i < ARR_SZ(CODEC_Config_Beep); i++)
    Codec_Write (CODEC_Config_Beep[i].Addr,
      CODEC_Config_Beep[i].Val);
} /* configureCodec */
```

13. Include this macro definition to calculate the size of a (const) array, as follows:

```
/* Calculate array size */
#define ARR_SZ(x) (sizeof (x) / sizeof(x[0]))
```

14. Define a global array of codec register address/value pairs named `CODEC_Config_Init`:

```
/***
 * CODEC initialization based on p38
 * of CS42L52 data sheet DS680F2
 *****/
REG_VAL CODEC_Config_Init[] = {
  {0x00, 0x99},
  {0x3E, 0xBA},
  {0x47, 0x80},
  {0x32, 0x80},
  {0x32, 0x00},
  {0x00, 0x00},
};
```

15. Define a global array of codec register address/value pairs named
    CODEC_Config_Beep:

```
/***
* CODEC initialization for Beep Generator
* of CS42L52 (Grant Ashton)
*****/
REG_VAL CODEC_Config_Beep[] ={
   /* Set I2S Ser. Mstr Op Only, for Beep Gen */
   {CS42L52_IFACE_CTL1, 0x80},
   /* Speaker Vol B=A, MONO */
   {CS42L52_PB_CTL2, 0x0A},
   /* Set master vol for A */
   {CS42L52_MASTERA_VOL, 0xC0},
   /* Ignore jpr setting */
   {CS42L52_PWRCTL3, 0xAA}
};
```

16. Create a new file named CS42L52.h defining symbolic names (for example,
    CS42L52_IFACE_CTL1, CS42L52_PB_CTL2, CS42L52_MASTERA_VOL, and so on)
    for CS42L52 register addresses. For example, as in the following addresses:

```
/* Register addresses */
#define CS42L52_CHIP_ID     0x01
#define CS42L52_PWRCTL1     0x02
#define CS42L52_PWRCTL2     0x03
#define CS42L52_PWRCTL3     0x04
#define CS42L52_CLK_CTL     0x05
// etc.
```

17. Add a function named genMCLK() to the codec_CS42L52.c file:

```
static void genMCLK(void) {
  GPIO_InitTypeDef GPIO_InitStruct;

  TIM3_Initialize();
  __GPIOC_CLK_ENABLE();

  /* Configure GPIO pin: PC6 */
  GPIO_InitStruct.Pin   = GPIO_PIN_6;
  GPIO_InitStruct.Mode = GPIO_MODE_AF_PP;
  GPIO_InitStruct.Pull  = GPIO_PULLUP;
  GPIO_InitStruct.Speed = GPIO_SPEED_FAST;
  GPIO_InitStruct.Alternate = GPIO_AF2_TIM3;
  HAL_GPIO_Init(GPIOC, &GPIO_InitStruct);
}
```

18. Add a function named `codecInitialize()` to the `codec_CS42L52.c` file. Note that the code to configure the I2C bus is identical to the code in `Touch_Initialize()`:

```
int32_t codecInitialize() {
    int32_t status;
    /* Configure I2C */
    ptrI2C->Initialize (NULL);
    ptrI2C->PowerControl (ARM_POWER_FULL);
    ptrI2C->Control (ARM_I2C_BUS_SPEED,
                          ARM_I2C_BUS_SPEED_FAST);
    ptrI2C->Control (ARM_I2C_BUS_SPEED,

    /* Configure CODEC */
    configureCodec();
    genMCLK();

    /* CODEC Power up     */
    status = Codec_Write(CS42L52_PWRCTL1, 0x00);
    delay_ms(10); /* Wait 10ms */

    return status;
}
```

19. Add a function named `readCodecChipID()` to the `codec_CS42L52.c` file:

```
int32_t readCodecChipID(uint8_t *val) {
    int32_t status = Codec_Read(1, val);

    return status;
}
```

20. Add a function named `Beep()` to the `codec_CS42L52.c` file:

```
void Beep(noteInfo note ) {
    /* Beep off time 1.23s and volume 0dB */
    Codec_Write(CS42L52_BEEP_VOL, 0x00);
    /* Set beep note and    beep duration */
    Codec_Write(CS42L52_BEEP_FREQ,
                  note.pitch | note.duration);
    /* play single beep */
    Codec_Write(CS42L52_BEEP_TONE_CTL, 0x40);
    /* Disable beep */
    Codec_Write(CS42L52_BEEP_TONE_CTL, 0x00);
}
```

21. Create the `timer.h` header file containing the `timer.c` function prototypes.

22. Create the `codec_CS42L52.h` header file containing the `codec_CS42L52.c` function prototypes.

23. Define symbolic names for the pitch of notes in the `codec_CS42L52.h` file, for example, as in the following frequencies:

```
// Beep note frequency
#define A5 0x60
#define A6 0xD0
#define B5 0x70
#define B6 0xE0
// etc.
```

24. Define symbolic names for the duration of notes in the `codec_CS42L52.h` file, for example, as in the following:

```
#define TENTH_SECOND    0x00
#define HALF_SECOND     0x01
#define ONE_SECOND      0x02
// etc.
```

25. Extend the `main()` function by adding code to initialize the GLCD and Codec. Define a super-loop that outputs a beep every 0.5 seconds:

```
int main (void) {
  noteInfo note = {G5, 0x02};

  uint8_t codecID;
  char buffer[128];

  HAL_Init ();    /* Init Hardware Abstraction Layer */
  SystemClock_Config ();              /* Config Clocks */

  GLCD_Initialize();
  setDisplay();

  showStatus(CodecInitialize());
  showStatus(readCodecChipID(&codecID));
  sprintf(buffer, "Chip ID: 0x%x", codecID);
  GLCD_DrawString (1*16, 9*24, buffer);

  while (1) {
    Beep(note);                       /* Play the note */

    wait_delay(500);                        /* pause */
  } /* WHILE */
}
```

26. Add a function named `setDisplay()` (copy the first 12 lines of the similarly-named function used in `touchScreenDemo_c6v0`).

27. Add a function named `showStatus( )`:

```
void showStatus(int32_t stat) {
   if (stat==0) GLCD_DrawString (1*16, 8*24,"Codec OK  ");
   else GLCD_DrawString (1*16, 8*24,"Codec FAIL");
}
```

28. Check that the `codec_CS42L52.c` and `timer.c` files are added to the project.

29. Select the **Use MicroLIB** project option.

30. Remember to configure the `RTE_Device.h` and `RTX_Conf_CM.c` files, as we did for the `touchScreenDemo_c6v0` folder from the *Setting the RTE for the I2C Peripheral Bus* recipe.

31. Build the project, then download and run the program.

How it works...

A Linux driver for the CS42L52 device has been written by Cirrus Logic (`http://lxr.free-electrons.com/source/sound/soc/codecs/cs42l52.c`) and is freely distributed under the terms of the **GNU General Public License**. So, we can use this together with information from the datasheet (`http://www.cirrus.com`) as a basis for our driver for the MCBSTM32F400 evaluation board. As the audio codec is also connected to the I2C serial bus, the touchscreen driver that we met in the previous section provides a good template for our audio codec driver. Therefore, we will organize the codec driver in three files that mirror those of the touchscreen driver, as follows:

 ► **CS42L52.h**: This defines codec registers

 ► **Codec_CS42L52.c**: This declares functions

 ► **Codec_CS42L52.h**: This declares function prototypes and defines symbolic names for constants

The code in the `Codec_CS42L52.c` file first defines the I2C port that is used to communicate with the audio codec. The board schematic confirms that the touchscreen and the audio codec are connected to the same I2C port (that is, serial clock SCL = PB8 and SDA = PB9), so we configure the RTE and RTX exactly as `touchScreenDemo_c6v0` using I2C port 1 (I2C1). The following preprocessor directives define the port number:

```
#ifndef CODEC_I2C_PORT
#define CODEC_I2C_PORT    1  /* I2C Port number*/
#endif
```

The following preprocessor macro ensures that the `ptrI2C` identifier points to the appropriate I2C driver:

```
/* I2C Driver */
#define _I2C_Driver_(n)   Driver_I2C##n
#define  I2C_Driver_(n) _I2C_Driver_(n)
extern ARM_DRIVER_I2C    I2C_Driver_(CODEC_I2C_PORT);
#define ptrI2C           (&I2C_Driver_(CODEC_I2C_PORT))
```

The most-significant 6-bit audio codec's I2C address is shown on the board schematic and the CS42L52 datasheet as 1001012. Bit-0 reflects the logic level of the AD0 pin (that is, 0 V), and the LSB is 0 (for write operations). So, our codec's I2C address is 0x94, that is, the following:

```
#define CODEC_I2C_ADDR    0x4A /* I2C address */
```

Note that in practice, all accesses to the codec are writes because the read protocol uses an abortive write cycle first to select the codec register before reading its contents (refer to `http://www.cirrus.com` for further details).

We declare two functions: `Codec_Write( )` and `Codec_Read( )`, which mirror `Touch_Write( )` and `Touch_Read( )`, which were declared in `Touch.c` to read and write to the audio codec.

The function named `CodecInitialize()` performs three tasks. It configures the I2C interface, then it generates the 12 MHz master clock MCLK (codec Pin 37), and finally, it performs the codec's initialization sequence.

The function named `genMCLK()` configures TIM3 to generate a 12-MHz clock and maps this onto the Alternate Function (AF) GPIO Port C pin 6 output. The initialization for TIM3 is similar to that described in the previous chapter except that we use the PWM mode with the capture/compare register to give an approximate 50% duty cycle. The code to configure the GPIO pin that is used to source MCLK is similar to the one that we saw in the `LED_Initialize()` function.

The initialization sequence for the audio codec is given on page 38 of the CS42L52 data sheet. The initialization sequence is stored in an array named `CODEC_RegInit[]`. The array entries are structured as follows:

```
/* Register value */
typedefstruct {
  uint8_tAddr;
  uint8_t Val;
} REG_VAL;
```

The register names (for example, `MASTERA_VOL`, and so on) are defined in the `CS42L52.h` header file (note that the register names can be copied from the Linux CS42L52 driver). To prevent odd pops and crackles, the data sheet advises that the chip is powered down before initialization and then powered up. This configuration code is included in the `configureCodec()` function. This function includes a nice example of a macro named ARR_ SZ to compute the size of the array:

```
/* Calculate array size */
#define ARR_SZ(x)  (sizeof (x) / sizeof(x[0]))
```

Note that unlike some languages, such as Java, C doesn't perform any array bounds checking, so it can be quite difficult to track errors due to incorrect array access; because of this, this macro is particularly useful.

In this recipe, we're only using the codec's beep generator (*section 4.3* of the data sheet), and the values stored in the `CODEC_Config_Beep[]` array are concerned with setting the codec up for this task. The remaining functions declared in the `codec_CS42L52.c` file are concerned with generating beeps and adjusting the volume of the speaker. The beep generator can be configured to produce single, multiple, or continuous beeps, but we only need single beeps to play our tune. The `Beep( )` function generates a single beep. This function takes an input parameter that determines the pitch and duration of the beep, and this is combined into one byte and written to the codec register address offset 0x1C in the format shown in the following table:

Bit-7	Bit-6	Bit-5	Bit-4	Bit-3	Bit-2	Bit-1	Bit-0
FREQ3	FREQ2	FREQ1	FREQ0	ONTIME3	ONTIME2	ONTIME1	ONTIME0

How to use the audio codec

Listening to the beep generated by codecDemo_c6v0 gets very annoying after a couple of minutes, so we will try and improve matters by adding a couple of functions that will enable us to change and mute the volume. We'll also modify the code to use the beep generator to play a tune. We're limited to a fairly simple tune because the beep generator only generates audio frequencies across two octave major scales. For those who are musically minded, we define the mapping between notes (pitch) and beep frequencies, and the beep ON time (see *section 6.21* of the data sheet) as well, in the `codec_CS42L52.h` header file. We call this recipe codecDemo_c6v1.

How to do it...

Follow the outlined steps to use the audio codec:

1. Clone the previous recipe and name the folder `codecDemo_C6v1`.

2. Open the RTE manager and add **Board Support** for **Buttons (API)** and **LED (API)**. Click **Resolve** and **OK**.

3. Add a function named `setVolume()` to the `codec_CS42L52.c` file:

```
static void setVolume(int32_t vol) {

    if (vol < -128)
        Codec_Write(CS42L52_MASTERA_VOL, (uint8_t) vol+256);
    else
        Codec_Write(CS42L52_MASTERA_VOL, (uint8_t) vol);
}
```

4. Add a function named `getVolume()` to the `codec_CS42L52.c` file:

```
int32_t getVolume( ) {
    int32_t vol, out_vol;
    uint8_t val;

    Codec_Read(CS42L52_MASTERA_VOL, &val);
    vol = (int32_t) val;
    if (vol > 24) {
        out_vol = -204; /* -102 db (saturated) */
        if (vol > 52) out_vol = vol-256;
    }
    else out_vol = vol;

    return out_vol;
}
```

5. Add a function named `decreaseVolume()` to the `codec_CS42L52.c` file:

```
void decreaseVolume(uint32_t stepSize) {
    int32_t currentVolume = getVolume();
    const int32_t minVolume = MIN_VOL_DB*2; /* -102dB */
    uint32_t n = 0;

    while ((currentVolume > minVolume) && (n<stepSize)) {
        currentVolume--; /* 0.5dB decrement */
        setVolume(currentVolume);
        n++;
    }
}
```

6. Add a function named `increaseVolume()` to the `codec_CS42L52.c` file:

```
void increaseVolume(uint32_t stepSize)
{
    int32_t currentVolume = getVolume();
    const int32_t maxVolume = MAX_VOL_DB*2;  /* +12dB */
    uint32_t n=0;

    while ((currentVolume < maxVolume) && (n<stepSize)){
        currentVolume++; /* 0.5dB increment */
        setVolume(currentVolume);
        n++;
    }
}
```

7. Add a function named `setMute()` to the `codec_CS42L52.c` file:

```
void setMute(bool state) {
    uint8_t val;

    if (state) val = 0x01;
    else val = 0x00;
    Codec_Write(CS42L52_PB_CTL1, val);
}
```

8. Declare a global constant array in the `codecDemo.c` file and assign values representing the notes for our tune:

```
noteInfo tune[] = {
    {G5, 0x02}, {G5, 0x02}, {A5, 0x02}, {F5, 0x04},
    {G5, 0x01}, {A5, 0x02}, {B5, 0x02}, {B5, 0x02},
    {C6, 0x02}, {B5, 0x04}, {A5, 0x01}, {G5, 0x02},
    {A5, 0x02}, {G5, 0x02}, {F5, 0x02}, {G5, 0x02},
    {G5, 0x01}, {A5, 0x01}, {B5, 0x01}, {C6, 0x01},
    {D6, 0x02}, {D6, 0x02}, {D6, 0x02}, {D6, 0x04},
    {C6, 0x01}, {B5, 0x02}, {C6, 0x02}, {C6, 0x02},
    {C6, 0x02}, {C6, 0x04}, {B5, 0x01}, {A5, 0x02},
    {B5, 0x02}, {C6, 0x01}, {B5, 0x01}, {A5, 0x01},
    {G5, 0x01}, {B5, 0x04}, {C6, 0x01}, {D6, 0x02},
    {E6, 0x01}, {C6, 0x01}, {B5, 0x02}, {A5, 0x02},
    {G5, 0x09} };
```

9. Replace function named `setDisplay()` in the `codecDemo.c` file:

```
void setDisplay( ) {

    GLCD_SetBackgroundColor (GLCD_COLOR_WHITE);
    GLCD_ClearScreen ();
```

```
GLCD_SetFont (&GLCD_Font_16x24);
GLCD_SetForegroundColor (GLCD_COLOR_BLACK);
GLCD_DrawString (1*16, 1*24, "Volume: ");
GLCD_DrawString (1*16, 5*24, "Wakeup toggles MUTE");
GLCD_DrawString (1*16, 6*24, "User and Tamper");
GLCD_DrawString (1*16, 7*24, "Adjust Volume");

#ifdef __DEBUG
showCodecInfo( );
#endif
}
```

10. Add a function named `volumeUserInput()` to the `codecDemo.c` file:

```
void volumeUserInput( ) {
  uint32_t keyMsk;

  keyMsk = Buttons_GetState ();
    if (keyMsk & BUTTONS_TAMPER_MASK)
      increaseVolume(10);
    else {
      if ( keyMsk & BUTTONS_USER_MASK )
        decreaseVolume(10);
      else
        if (keyMsk & BUTTONS_WAKEUP_MASK) {
          mute = !mute;
          setMute(mute);
    } /* IF-ELSE */
  } /* IF-ELSE */
}
```

11. Add a function named `showVolumeGraph()` to the `codecDemo.c` file:

```
void showVolumeGraph( ) {

  if (mute) {/* If codec is muted, display red graph */
    GLCD_SetForegroundColor (GLCD_COLOR_RED);
    GLCD_DrawString(1*16, 2*24, "(Muted)");
  }
  else {                        /* else blue graph */
    GLCD_SetForegroundColor (GLCD_COLOR_BLUE);
    GLCD_DrawString(1*16, 2*24, "          ");
  }
  GLCD_DrawBargraph(130, 24, 180, 20,

                    (getVolume() - (MIN_VOL_DB*2))/2);
}
```

12. Replace the `main()` function in the `codecDemo.c` file:

```c
int main (void) {

    uint32_t i = 0;
    uint32_t beepTimeOut = 0;

    HAL_Init ();    /* Init Hardware Abstraction Layer */
    SystemClock_Config ();              /* Config Clocks */

    GLCD_Initialize();
    LED_Initialize ();
    Buttons_Initialize ();
    CodecInitialize();
    setDisplay( );

    while (1) {
        if (!beepTimeOut) {
            Beep(tune[i]);                 /* Play the next note */
            beepTimeOut = tune[i].duration;
            i = (i+1)%ARR_SZ(tune);
        }
        else
            beepTimeOut--; /* Wait */

        volumeUserInput ( );
        showVolumeGraph ( );
        LED_SetOut (i);

        wait_delay(BEAT_TIME);
    } /* WHILE */
}
```

13. Build, download, and run the program. You should get something similar to the following screenshot

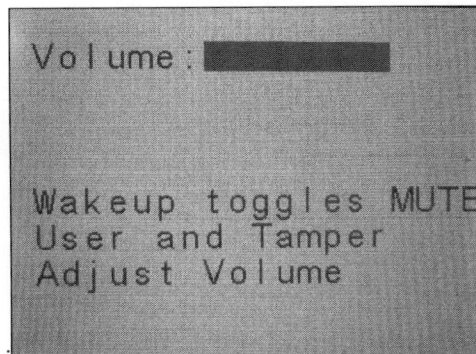

How it works...

The functions in steps 2 and 3 of this recipe are concerned with controlling the speaker volume. The `setVolume()` function can be made `static` to enforce privacy (static functions can only be called within the file in which they are defined). Both functions access the register that controls the master volume for codec channel A (Address Offset 0x20):

Bit-7	Bit-6	Bit-5	Bit-4	Bit-3	Bit-2	Bit-1	Bit-0
MSTxVOL7	MSTxVOL6	MSTxVOL5	MSTxVOL4	MSTxVOL3	MSTxVOL2	MSTxVOL1	MSTxVOL0

The master the volume is represented using a special 8-bit 2's-complement code, which allocates values 0-24 to positive numbers, and the remaining values to negative ones (note that a normal 8-bit 2s-complement representation allocates code words equally between positive and negative quantities). The function to read the `getVolume( )` master volume register converts the value read from the register to a signed 32-bit integer that represents twice the volume in decibels (dB) (that is, values between -204 and +24 represent -102 dB to +12 dB).

An input parameter of the `setVolume()` function represents twice the volume (dB). If the volume lies in the -128 to +24 range, then it simply casts the 32-bit signed integer as an 8-bit value before writing it to the codec's register. Otherwise, it adds an offset of +256. The binary code that is used to represent the volume is explained in the CS42L52 data sheet. The `inceaseVolume()`, `decreaseVolume()`, and `setMute()` functions described in steps 4 to 6 of the recipe provided a simple high-level interface that allows the volume to be manipulated.

Now that we have defined a codec driver, we can turn our attention to writing the main function for our audio codec demo. This simply needs to initialize the codec and then write appropriate values to the beep generator. The pitch and time values are stored in the global array named `tune [ ]`; can you guess the 'tune'? The **wakeup**, **tamper**, and **user** buttons are used to increase, decrease, and mute the volume, so they need to be initialized too. The super-loop inside `main()` outputs the array values (notes and durations by stepping through the `tune [ ]` array. The `VolumeUserInput()` function checks and processes button inputs, and the `ShowVolumeGraph()` function displays a bar indicating the volume on the GLCD. The function named `wait_delay()` ensures that each call to `Beep( )` is separated by an appropriate time interval set by the `BEAT_TIME` constant.

How to use the camera

The camera is another I2C peripheral, but to display video we need to read the array pixels that make up an image and write their values to the GLCD very rapidly. We achieve this by using **Direct Memory Access** (**DMA**) to stream image frames directly to the GLCD rather than writing individual values as we did for the audio codec demo. We'll name this recipe `cameraDemo_c6v0`.

How to do it...

1. Create a new project named `cameraDemo`. Using the RTE manager, go to **Board Support** and select the **Camera (API)** and **Graphic LCD (API)** software components.

2. Set the **CMSIS** and **Device** software components, as we've done for previous projects. Set the **Use MicroLIB** project option.

3. Create a file named `cameraDemo.c` and add boilerplate code to configure clocks, and so on. Add this file to the project.

4. Add a `main()` function to the `cameraDemo.c` file:

5. Build, download, and run the program, as follows:

```c
int main (void) {

  uint32_t addr;

  HAL_Init();          /* Initialize the HAL Library */
  SystemClock_Config();       /* Config System Clk */
  GLCD_Initialize();      /* Graphical Display Init */

                    /* Get fremebuffer addr */
  addr = GLCD_FrameBufferAddress();
  Camera_Initialize(addr);          /* Camera Init */


    /* Prepare display for video stream from camera */
  GLCD_SetBackgroundColor (GLCD_COLOR_BLUE);
  GLCD_ClearScreen ();
  GLCD_FrameBufferAccess (true);
  /* Turn camera on */
  Camera_On ();

  while (1) {
    ;        /* Nothing to do here; all done by DMA */
  }
}
```

6. Open the `RTE_Device.h` file and use the configuration wizard to set the I2C port parameters. Remember to check the DMA transmit and receive options (we can accept the default DMA parameters):

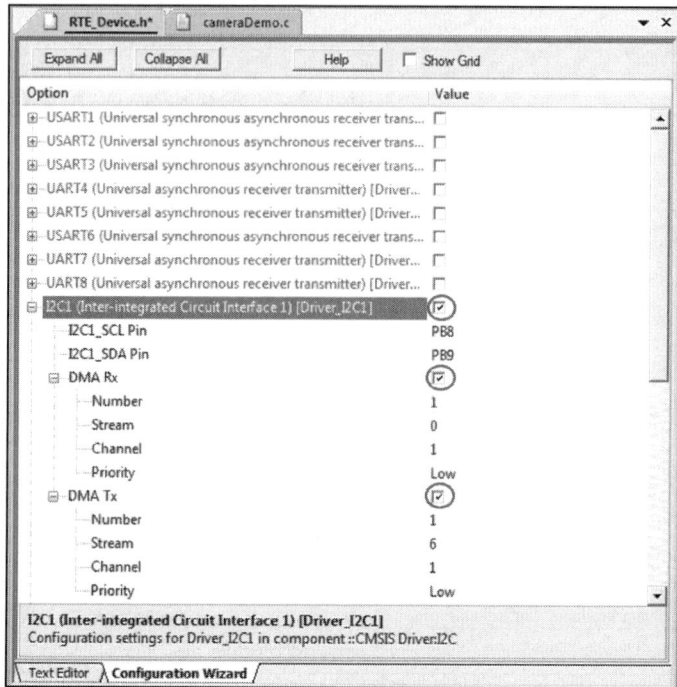

7. Build the project, then download and run the program.

How it works...

As the camera is another I2C peripheral and the driver (API) named `Camera_OVM7690.c` provided by ARM is structured in a similar way to that for the touchscreen and audio codec, the array named `Camera_RegInit [ ]` stores a number of addresses and value pairs that are written by the function named `Camera_Initialize()`. The camera used on the evaluation board is an OVM7690 part manufactured by OmniVision (`http://www.ovt.com`). The camera resolution is 640 × 480 pixels and operates at up to 30 frames per second (fps). We need to access OmniVision's OVM7690 Software Application Note in order to understand the code used to initialize the camera, but currently, these documents are company-confidential and protected by Non-Disclosure Agreements (NDAs). The camera is aimed at mobile phone, notebook, and automotive applications and includes a number of programmable controls for image-processing functions, such as exposure, gamma, white balance, hue, and so on. `Camera_Initialize()` also configures a DMA channel to stream data from the camera to SDRAM, so it needs to be provided with the base address of a memory segment. This address is defined by the GLCD (API) and acquired by the `GLCD_FrameBufferAddress ( )` function.

There's more...

A demo project that exercises many of the features of the MCBSTM32F400 evaluation board can be downloaded by the pack installer with the Device Family Pack (currently the version is DFP 2.6.0). As the demo program displays icons on the GLCD that are encoded as bitmaps, the executable image for the program exceeds the limit imposed by the evaluation version of the uVision IDE. This code is read-only, but it has been precompiled so that the project can be downloaded and run on the board.

The main function declared in the demo.c file implements a **finite-state machine** (**FSM**) that determines the operating mode of the program. An integer variable named mode is assigned a value of 0, 1, 2, or 3 depending on the mode that was selected. These modes are mapped to the M_INIT, M_STD, M_MOTION, and M_CAM literals by the enumerated type definition:

```
/* Mode definitions */
enum {
   M_INIT = 0,
M_STD,
   M_MOTION,
   M_CAM,
};
```

The mode variable is assigned by the function called SwitchMode() that takes an input argument that identifies the current state (that is, 0, 1, 2, and 3) and returns the next state. For example, the first call to SwitchMode() is made when the current state is M_INIT:

```
mode = SwitchMode (M_INIT);
```

A switch statement in main() determines different behaviors for each mode, as follows:

```
switch (mode) {
case M_STD:
    ...
break;
case M_MOTION:
    ...
break;
case M_CAM:
    ...
break;
default:
mode = SwitchMode (mode);
break;
}
```

This behavior is better described by a state diagram (shown as follows). This diagram is a graph where states are identified by vertices and the permitted transitions between states by edges. The edges are labeled with events that give rise to the changes of state.

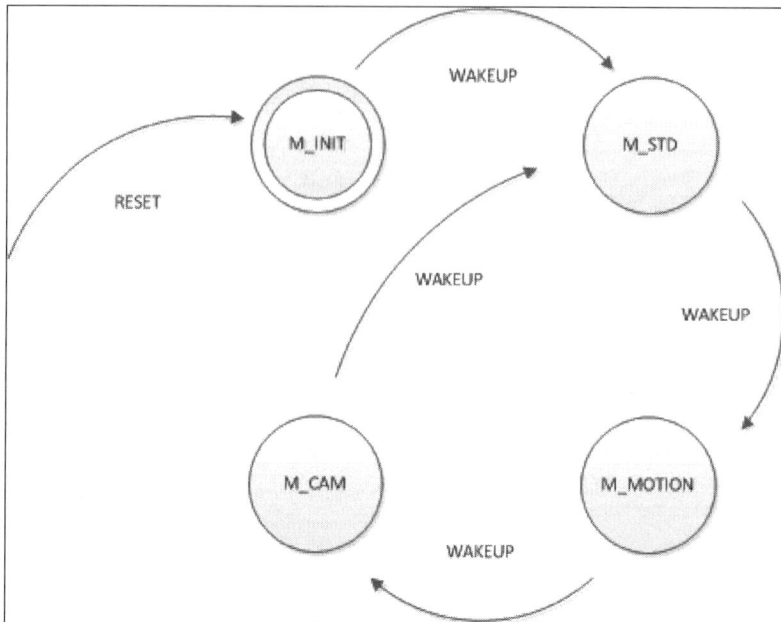

The Demo project is a very useful resource as it provides example code for many of the evaluation board functions. The #include statements at the start of the main source file provide some insight into what is available:

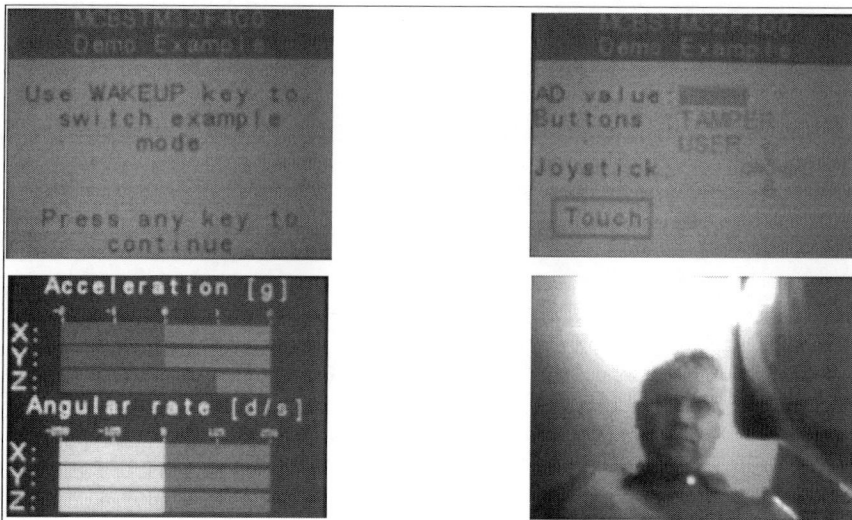

Designing bitmapped graphics

User interfaces and games can be made much more interesting using color graphics. The GLCD library includes a function called GLCD_DrawBitmap() that can be used to render 16-bit color bitmaps. Bitmaps can be designed using standard editors or downloaded from elsewhere. The following recipe shows you how to generate a simple bitmapped representation of a ball that can be used with the *helloBounce* and *helloPong* recipes we developed in *Chapter 2, C Language Programming*. We'll call this recipe bitmapBounce_c6v0.

How to do it...

To design bitmapped graphics, follow these instructions:

1. Create a color bitmap of width 16 pixels and height 24 pixels using the Windows Paint application. A screenshot of what this should look like is displayed, as follows:

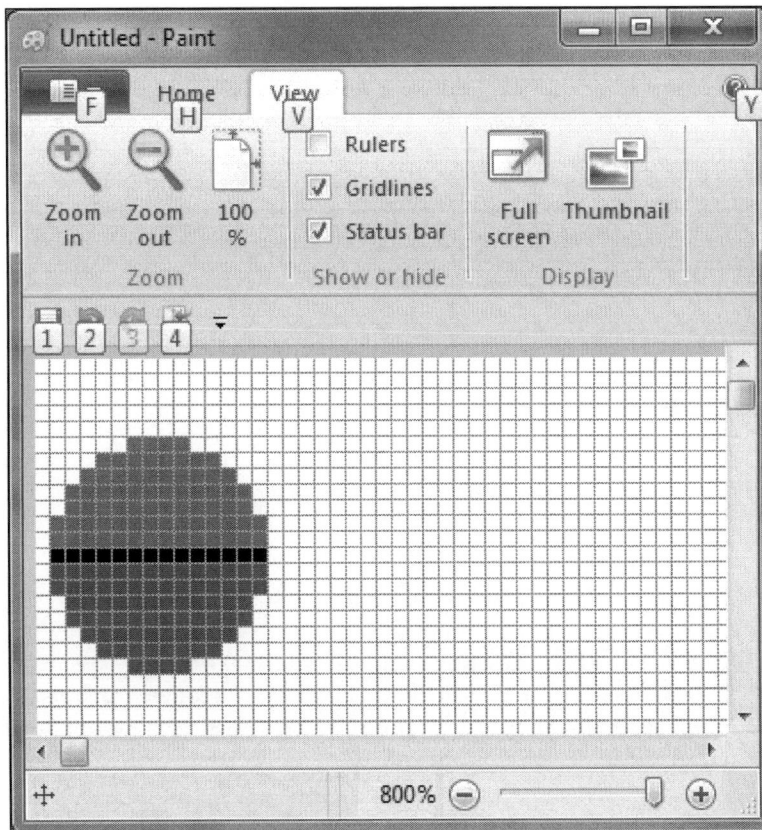

2. Save the ball icon as a standard 24-bit bitmap, with the filename as ball.bmp.

3. Use GIMP (`http://www.gimp.org`) to convert the 24-bit-per-pixel bitmap to a 16-bit-per-pixel format and store the pixel values in an array. First install GIMP and open the `ball.bmp` file.

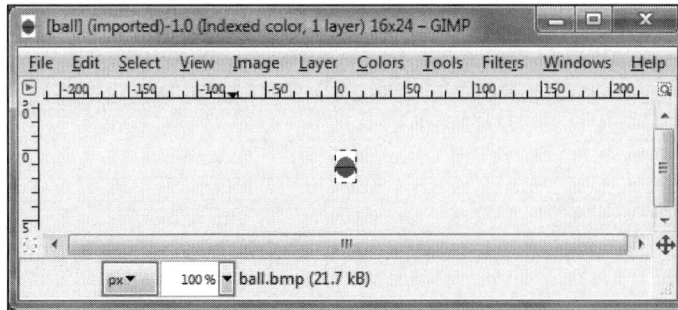

4. Export the image as a C source file in 16-bit format using the GIMP export sub-menu. This creates the C source file (in this case, named `ball_16bit.c`).

5. Clone the folder named `helloBounce_c2v0` from the *Creating a game application – Stage 1* recipe that we introduced in *Chapter 2, C Language Programming*, and cut and paste the contents of the `ball_16bit.c` file into `helloBounce.c`, as follows:

```
/* GIMP RGB C-Source image dump (ball_16bit.c) */
static const struct {
unsigned int    width;
unsigned int    height;
unsigned int    bytes_per_pixel; /* 2:RGB16 3:RGB 4:RGBA */
```

```
unsigned char      pixel_data[16 * 24 * 2 + 1];
} gimp_image = {
   16, 24, 2,
   "\377\377\377\377\377\377\... etc.
}
```

6. Delete the `extern GLCD_FONT      GLCD_Font_16x24;` declaration:

7. Search for the following references:

```
GLCD_Font_16x24.width
GLCD_Font_16x24.height
```

Replace these reference with the following ones:

```
gimp_image.width
gimp_image.height
```

8. Delete the call to `GLCD_SetFont (&GLCD_Font_16x24);`.

9. Search for the following statement:

```
GLCD_DrawChar ( x, y, 0x81 );
```

Replace this statement with the following one:

```
GLCD_DrawBitmap ( x, y, gimp_image.width,
         gimp_image.height, gimp_image.pixel_data );
```

10. Rebuild, download, and run the program.

How it works...

The ball used in the original recipes in *Chapter 2, C Language Programming*, is rendered using the filled circle character, which is one of a number of binary character bitmaps defined in a file named `GLCD_Fonts.c`. We're now using the `GLCD_Bitmap()` function to render the ball rather than `GLCD_DrawChar()`. This function expects a pointer to a 16-bpp bitmap. The bitmap data is provided by GIMP. The escape sequences \377\377\377, and so on, represent characters in the string encoded in octal. Therefore, 3778 = 111111112 and two bytes encode each 16-bit pixel, so 16-bit bitmaps can represent 65,536 colors. If the alpha channel is omitted (as in our case), then RGB channels are encoded by 5, 6, and 5-bits, respectively.

There's more...

The pixel data field of `gimp_image` comprises 16 x 24 x 2 + 1 = 769 bytes. If we store larger images in this way, our executable code image will quickly exceed the maximum allowed under the terms of our free MDK license. However, after examining the values in the array, we can see that many of the values are repeated, and this suggests that there may be a more efficient way of storing the pixel values. **Run-length encoding** (**RLE**) is a lossless compression algorithm that exploits the fact that there are often many repeated values in a bitmap (that is, adjacent pixels are often the same color). There are many variations of run length encoding, and a good introduction to the topic is given by Arturo Campos (`http://www.arturocampos.com/ac_rle.html`). We can export a run length encoded-version of our 16-bit BMP using GIMP.

GIMP adopts a run length encoding format known as **PackBits**, which was originally developed by Apple. A data stream encoded by PackBits comprises a series of packets. Each packet consists of a one byte header followed by data. The header byte (*n*) is interpreted as a signed value (8-bit 2's complement) and the data. A positive value (*n*) indicates that the *n* data elements that follow should be interpreted as literal values, and a negative value implies that the single data element that follows should be repeated *n* times. The data structure (produced by GIMP) with run length encoded data representing the pixel values exported from the `ball.bmp` file is as follows:

```
/* GIMP RGB C-Source image dump 1-byte-run-length-encoded
   (ball_16-bit_rle.c) */
static const struct {
unsigned int  width;
unsigned int  height;
```

```
unsigned int  bytes_per_pixel; /* 2:RGB16 3:RGB 4:RGBA */
unsigned char  rle_pixel_data[390 + 1];
  } gimp_image = {
    16, 24, 2,
    "\325\377\377\5\377\377\... etc.
}
```

The run length encoded image comprises just 391 bytes (approximately 50% compression).To render the encoded bitmap, we'll need to define a version of GLCD_Bitmap() that unpacks the data before writing it to the GLCD:

```
int32_t GLCD_RLE_Bitmap (uint32_t x, uint32_t y, uint32_t width,
uint32_t height, const uint8_t *bitmap) {

  int32_t npix = width * height;
  int32_t i=0, j;
  uint16_t *ptr_bmp;
  uint8_t count;

#if (GLCD_SWAP_XY == 0)
  y = (y + Scroll) % GLCD_HEIGHT;
#endif

  GLCD_SetWindow(x, y, width, height);

  wr_cmd(0x22);
  wr_dat_start();

  while (i<npix) {
    count = *bitmap++;
    ptr_bmp = (unsigned short *) bitmap;

    if (count >= 128) {
      count = count-128;
      for (j = 0; j<count; j++) { /* repeated pixels */
      wr_dat_only(*ptr_bmp);
    }
    bitmap+=2; /* adjust the pointer */
  }
  else {
    for (j=0; j<count; j++)
    wr_dat_only(ptr_bmp[j]);
    bitmap+=(count*2); /* adjust the pointer */
  }
  i+=count;
  } /* while */

  wr_dat_stop();
  return 0;
}
```

As the library source file, GLCD_MCBSTM32F400.c, is read-only, we'll need to add the GLCD_RLE_Bitmap() function to a local copy (named GLCD_MCBSTM32F400_plus.c). We'll also need to add a local copy of Board_GLCD.h (Board_GLCD_plus.h) that includes the function prototype, GLCD_RLE_Bitmap(). Remember to modify the conditional preprocessor statement, as follows:

```
#ifndef __BOARD_GLCD_PLUS_H
#define __BOARD_GLCD_PLUS_H
```

Include a modified version of the header in rle_bounce.c and GLCD_MCBSTM32F400_plus.c. We've named this recipe that uses run length encoding rleBounce_c6v0.

Ideas for games using sound and graphics

The scope to develop games for the MCBSTM32F400 evaluation board is unlimited; however, the restricted memory image imposed by the evaluation version of the MDK constrains their complexity and the size of bitmaps that can be used (we address this issue in *Chapter 9, Embedded Toolchain*). A number of general introductory texts on game development can inspire new ideas. While we used the topic of generating audio mainly to introduce the audio codec, it is a topic in its own right and those who wish to create a really professional gaming experience should refer to the book, The essential guide to game audio: The theory and practice of sound for games (http://www.taylorandfrancis.com/books). Screenshots of a few examples of games developed by students studying embedded systems are shown in the following screenshot:

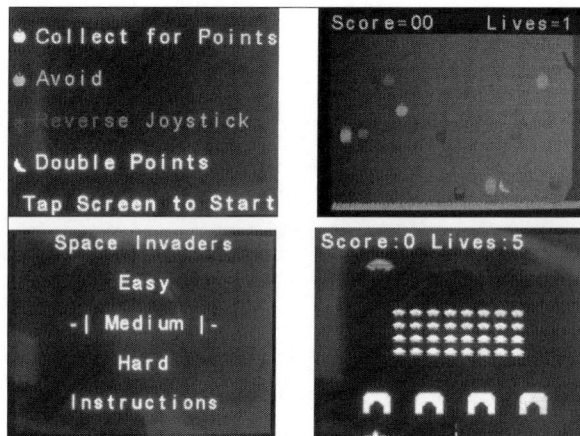

The board lends itself to single-player games but two-player scenarios can be accommodated by designing an Artificial Intelligence (AI) opponent. Two (human) players can compete either by taking turns or linking two boards together using the RS232 COM port.

7
Real-Time Signal Processing

In this chapter, we will cover the following topics:

- ▶ Configuring the audio codec
- ▶ How to play prerecorded audio
- ▶ Designing a low-pass digital filter
- ▶ How to make an audio tone control

Introduction

In the last chapter, we used the audio codec's beep generator to play a tune, but if you looked at the codec manufacturer's data sheet, you must have noticed that the device can do much more. Audio signals can be recorded by connecting a microphone to the evaluation board's stereo analog audio input, and the signal can be sampled using the audio codec's on-chip ADC. Digital audio can be played by sending digital samples to the codec, and the left and right speakers can be driven by the output of an on-chip DAC. A dedicated digital serial audio interface using a protocol called I2S (I2S, or IIS) conveys digital samples between the microcontroller and audio codec. Inter-IC-Sound (I2S) or **Integrated Interchip Sound (IIS)** is a serial bus interface standard developed by Phillips Semiconductors in 1986 (revised 1996) that is used to connect digital audio devices together. This specification is widely available online (for example, www.cypress.com). Unfortunately, the STM32F400 evaluation board only supports a half-duplex channel, so audio cannot be recorded and played simultaneously.

Connecting a powerful microcontroller (that is, the computer) and codec together brings the prospect of **Digital Signal Processing (DSP)**. DSP applications manipulate digital audio samples to create digital filters and other amazing audio effects.

Configuring the audio codec

The STM32F400 evaluation board schematic (http://www.keil.com) shows that a Cirrus Logic CS42L52 codec IC (http://www.cirrus.com) is used, and the I2S bus signals are driven by GPIO port I bits 0, 1, and 3. SDIN and SDOUT are wired together, so the I2S interface must be operated half-duplex. In addition to managing the I2S interface, the microcontroller must also source a **Master Clock** (**MCLK**), which clocks the codec's delta-sigma modulators (Note that we described a function to achieve this in *Chapter 6, Multimedia Support*). A block diagram that summarizes the I2S interface connection is shown, as follows:

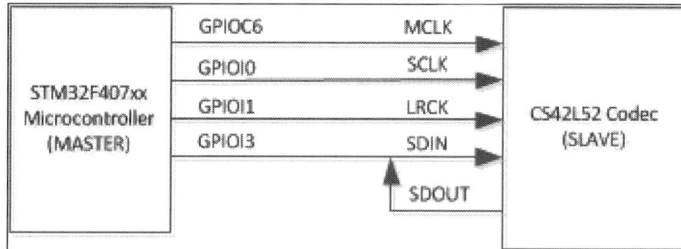

The codec also uses MCLK to power an inverter, which supplies a higher DC voltage to support analog parts of the codec. The codec data sheet explains that MCLK should be instantiated and the codec's registers must be configured while the device is powered down and the power up/down sequence outlined in the data sheet must be carefully followed to ensure the codec operates correctly.

The I2S specification identifies master and slave roles. An I2S bus must include one master (to source SCLK and LRCK), and it may include more than one slave. Normally, the microcontroller is configured as master, and as SDIN and SDOUT are connected together (externally), SDOUT must be switched to a high-impedance (HI-Z) state before SDIN is driven. If we refer to the following table the only option that allows for SDOUT to be HI-Z is to configure the codec as slave:

3ST_SP	Serial port status	
	Slave mode	Master mode
0	This is when serial port clocks are inputs, and SDOUT is output.	This is when serial port clocks and SDOUT are outputs.
1	This is when serial port clocks are inputs, and SDOUT is HI-Z.	This is when serial port clocks and SDOUT are HI-Z.

The microcontroller's **Serial Peripheral Interface** (**SPI**) and I2S interface is described in section 28 of STM's RM0090 Reference Manual (http://www.st.com). The following recipe, codecDemo_c7v0, describes how to configure the codec and output a continuous audio tone.

How to do it...

1. Clone `codecDemo_c6v0` from the *Writing a driver for the audio codec* recipe in *Chapter 6, Multimedia Support* to a new folder named `codecDemo_c7v0`.

2. Configure the Runtime Environment, as we did for the folder, `codecDemo_c6v0` from the *Writing a driver for the audio codec* recipe in *Chapter 6, Multimedia Support*, and add support for **Device → STM32Cube HAL → I2S**, as follows:

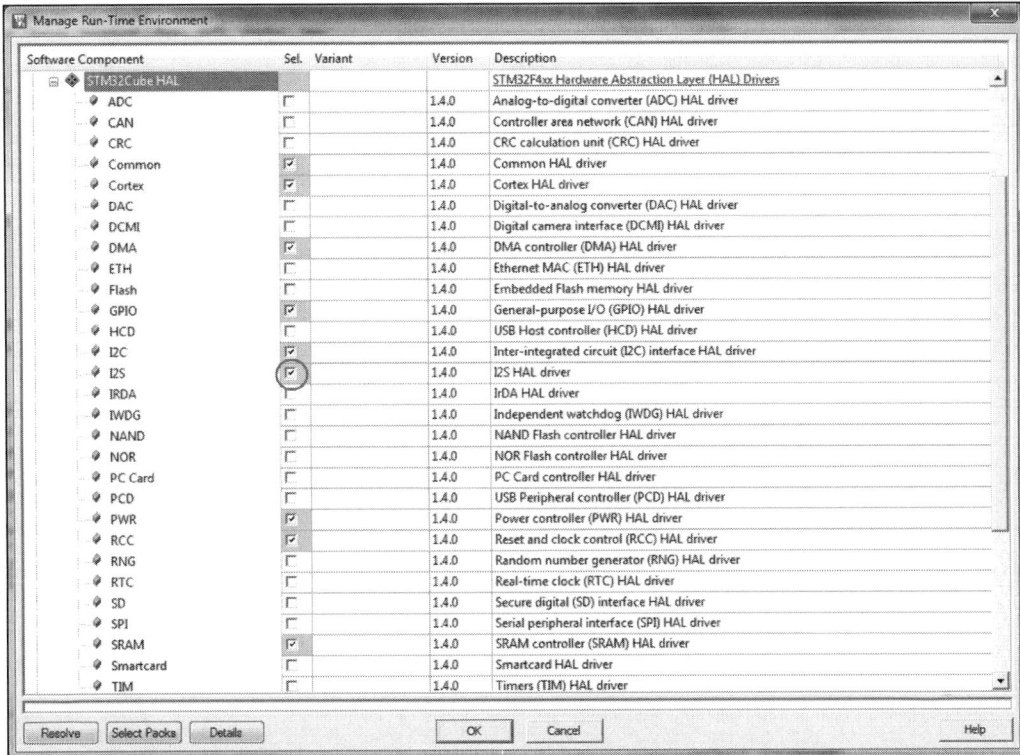

> There is no need to select **CMSIS Driver→ SPI (API)**.

3. Use the **Configuration Wizard** tabs in `RTE_Device.h` and `RTX_Conf_CM.c` to configure I2C and RTX parameters, as we did in the folder, `codecDemo_c6v0` from the *Writing a driver for the audio codec* recipe in *Chapter 6, Multimedia Support*.

4. Create a new file named `I2S_audio.c` and add this to the project:

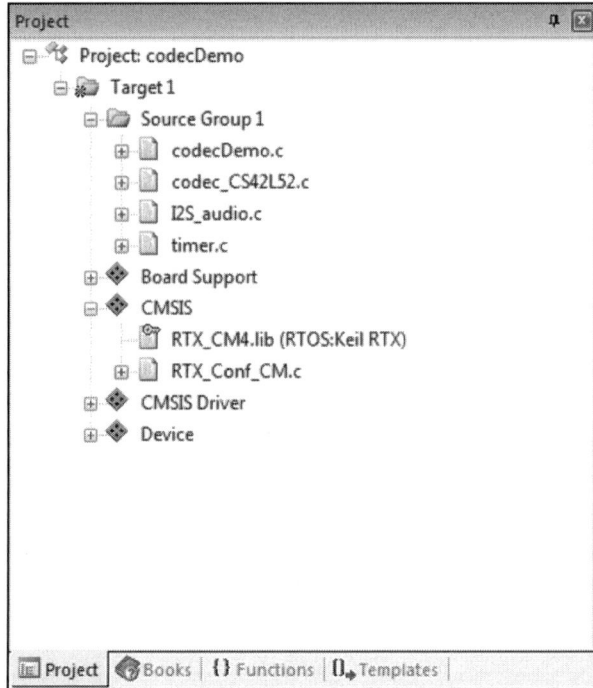

5. Add a global `I2S_HandleTypeDef` handle structure in the `I2S_audio.c` file, as follows:

```
/* Global I2S handle structure */
I2S_HandleTypeDef hi2s;
```

6. Define the `Set_I2S_GPIO_Pins()` function in the `I2S_audio.c` file, as follows:

```
void Set_I2S_GPIO_Pins(void) {
  GPIO_InitTypeDef GPIO_InitStruct;

  __GPIOC_CLK_ENABLE();
  __GPIOI_CLK_ENABLE();

    /* Configure GPIO pin: PI0,1,3 */
  GPIO_InitStruct.Pin     = GPIO_PIN_0 |
                          GPIO_PIN_1 | GPIO_PIN_3;
  GPIO_InitStruct.Mode = GPIO_MODE_AF_PP;
  GPIO_InitStruct.Pull  = GPIO_NOPULL;
  GPIO_InitStruct.Speed = GPIO_SPEED_FAST;
  GPIO_InitStruct.Alternate = GPIO_AF5_SPI2;
```

```
    HAL_GPIO_Init(GPIOI, &GPIO_InitStruct);

    /* Configure GPIO pin: PC6 */
    GPIO_InitStruct.Pin   = GPIO_PIN_6;
    GPIO_InitStruct.Mode = GPIO_MODE_AF_PP;
    GPIO_InitStruct.Pull  = GPIO_NOPULL;
    GPIO_InitStruct.Speed = GPIO_SPEED_FAST;
    GPIO_InitStruct.Alternate = GPIO_AF5_SPI2;
    HAL_GPIO_Init(GPIOC, &GPIO_InitStruct);
}
```

7. Define the `I2S_Audio_Initialize()` function (skeleton) in the `I2S_audio.c` file:

```
HAL_StatusTypeDef I2S_Audio_Initialize(void) {
  HAL_StatusTypeDef status;

  /* Enable the SPIx interface clock. */

  /* Configure I2S Pins */

  /* Program the Mode, Standard, Data Format,
     MCLK Output, Audio frequency and Polarity
     using HAL_I2S_Init() function. */
}
```

8. Add this code to enable the clock in the `I2S_Audio_Initialize()` function:

```
/* Enable the SPIx interface clock. */
RCC->CR |= RCC_CR_PLLI2SON;   /* Enable the PLLI2S */
                /* Wait till the main PLL is ready */
while((RCC->CR & RCC_CR_PLLI2SRDY) == 0)
    {}
__HAL_RCC_SPI2_CLK_ENABLE();
```

9. Call `Set_I2S_GPIO_Pins()`, as follows:

```
/* Configure I2S Pins */
Set_I2S_GPIO_Pins( );
```

10. Set the appropriate fields of the global `I2S_HandleTypeDef` handle structure and call `HAL_I2S_Init( )`:

```
/* Program the Mode, Standard, Data Format,
   MCLK Output, Audio frequency and Polarity
   using HAL_I2S_Init() function. */

  hi2s.Instance = SPI2;
  hi2s.State = HAL_I2S_STATE_RESET;
```

```
hi2s.Init.Mode = I2S_MODE_MASTER_TX;
hi2s.Init.Standard = I2S_STANDARD_MSB;
hi2s.Init.DataFormat = I2S_DATAFORMAT_16B;
hi2s.Init.MCLKOutput = I2S_MCLKOUTPUT_ENABLE;
hi2s.Init.AudioFreq = I2S_AUDIOFREQ_22K;
hi2s.Init.CPOL = I2S_CPOL_LOW;
hi2s.Init.ClockSource = I2S_CLOCK_PLL ;
hi2s.Init.FullDuplexMode = I2S_FULLDUPLEXMODE_DISABLE;

status = HAL_I2S_Init(&hi2s);
```

11. Add the `#include` files to the `I2S_audio.c` file:

```
#include "codec_CS42L52.h"
#include "stm32f4xx_hal.h"
#include "I2S_audio.h"
#include "stm32f4xx_hal_i2s.h"
```

12. Open the `codec_45L52.c` file and add an array of register or value pairs to configure the codec for sampled audio:

```
REG_VAL CODEC_Audio_I2S_Slave[] ={
  /****
   *Configure I2S Interface as Slave, 16bits
   ******/
  {CS42L52_IFACE_CTL1, 0x03},
  /* SDOUT is HI-Z */
  {CS42L52_IFACE_CTL2, 0x10},
  /* Speaker Vol B=A, MONO */
  {CS42L52_PB_CTL2, 0x0A},
  /* Set master vol for A/B */
  {CS42L52_MASTERA_VOL, 0xC0},
  /* Ignore jpr setting (speaker always ON) */
  {CS42L52_PWRCTL3, 0xAA}
};
```

13. Modify the function named `configureCodec ( )` so that we can select an appropriate setup, depending on an input argument named `mode`:

```
static void configureCodec(codecMode mode) {
  uint32_t i;

  Codec_Write(0x02, 0x01);   /* Keep Codec Power-down */
  delay_ms(10); /* Wait 10ms */

  for (i = 0; i < ARR_SZ(CODEC_Config_Init); i++)
    Codec_Write (CODEC_Config_Init[i].Addr,
```

```
                 CODEC_Config_Init[i].Val);

    if (mode == AUDIO_BEEP)
      for (i = 0; i < ARR_SZ(CODEC_Config_Beep); i++)
        Codec_Write (CODEC_Config_Beep[i].Addr,
                     CODEC_Config_Beep[i].Val);
    else
      if (mode == AUDIO_SAMPLED)
        for (i = 0; i < ARR_SZ(CODEC_Audio_I2S_Slave); i++)
          Codec_Write (CODEC_Audio_I2S_Slave[i].Addr,
                       CODEC_Audio_I2S_Slave[i].Val);


  }
```

14. Use `mode` to manage calls to `configureCodec()` and `genMCLK()` in the `codecInitialize()` function:

```
/* Configure CODEC */
  configureCodec(mode);

  /* Configure I2S */
  if (mode == AUDIO_SAMPLED)
    status = I2S_Audio_Initialize();
  else
    if (mode == AUDIO_BEEP)
      genMCLK();
```

15. Define mode in `codec_42L52.h`, as follows:

```
typedef enum {
  AUDIO_BEEP,
  AUDIO_SAMPLED
} codecMode;
```

16. Open the `codecDemo.c` file and add the following:

```
#include "I2S_audio.h"
#include "stm32f4xx_hal_i2s.h"

/* Timeout value fixed to 100 ms */
#define I2S_TX_TIMEOUT_VALUE ((uint32_t)100)
/* Macro to calculate array size */
#define ARR_SZ(x) (sizeof (x) / sizeof(x[0]))


/* Global External Vars */

extern I2S_HandleTypeDef hi2s;
```

17. Add a global `const` array of audio samples to the `codecDemo.c` file:

```
/* 20 left+right channel samples @ 22kHz ~= 1.4 kHz. */
const int16_t dacLUT [ ] = {
                        0,          0,      9830,       9830,     19660,
                    19660,     26214,     26214,      31456,     31456,
                    32767,     32767,     31456,      31456,     26214,
                    26214,     19660,     19660,       9830,      9830,
                        0,          0,     -9830,      -9830,   -19661,
                   -19661,    -26214,    -26214,     -31457,    -31457,
                   -32768,    -32768,    -31457,     -31457,    -26214,
                   -26214,    -19661,    -19661,      -9830,     -9830  };
```

18. Modify the `main()` function in the `codecDemo.c` file. Add and initialize the variable mode and pass the value to `CodecInitialize()`, as follows:

```
int main (void) {
   noteInfo note = {G5, 0x02};
   codecMode mode = AUDIO_SAMPLED;
   HAL_StatusTypeDef status;

   /* Uncomment for BEEP */
   //mode = AUDIO_BEEP;

   HAL_Init ( );
    SystemClock_Config( );
    GLCD_Initialize();
     status = CodecInitialize(mode);
    setDisplay( );

   // etc.

   }
```

19. If required, we can add a function named `showCodecI2SInfo()` that displays the status (to debug):

```
#ifdef __DEBUG
   if (mode == AUDIO_SAMPLED)
      showCodecI2SInfo(status);
#endif
```

20. Modify the super loop in `main()` and call `HAL_I2S_Transmit()`, as follows:

```
while (1) {
   if (mode == AUDIO_BEEP) {
      Beep(note);                        /* Play the note */
      wait_delay(500);                   /* pause */
```

```
   }
   else
      if (mode == AUDIO_SAMPLED)      /* Play a tone */
         HAL_I2S_Transmit(&hi2s, (uint16_t *) dacLUT,
                   ARR_SZ(dacLUT), I2S_TX_TIMEOUT_VALUE );
} /* WHILE */
```

21. Uncomment the `mode = AUDIO_BEEP;` statement. Build and run the program to confirm that I2C communication with the audio codec is established and the program performs as `codecDemo_c6v0` from the *Writing a driver for the audio codec* recipe in *Chapter 6, Multimedia Support*.

22. Reinstate the comment. Build, download, and run the code. We should now hear a shrill tone.

How it works...

Before powering the codec up (by clearing bit 0 of the codec's power control 1 register), we must first ensure that MCLK is established. As we're using the `stm32f4xx_hal_i2s.h` HAL library to manage the I2S low-level interface, we can take advantage of its ability to generate MCLK rather than configuring a timer as we did in *Chapter 6, Multimedia Support*. The I2S bus and audio codec channels are configured by a function named `I2S_Audio_Initialize()`, which, in turn, is called by `CodecInitialize()`. The `I2S_Audio_Initialize()` function performs the tasks that are identified in the comment at the start of the `stm32f4xx_hal_i2s.c` file. This enables the I2S clock, configures the GPIO pins, sets GPIO for I2S **Alternate Function (AF)**, sets the I2S handle struct, and initializes the I2S peripheral (using the HAL device driver). Referring to STM's reference manual, RM0090 (http://www.st.com), we can see that the microcontroller has a number of I2C and SPI peripherals, which begs the question, *How do we decide which instance of a peripheral to use?* The answer is that, as we're using an evaluation board, the board's designer already made this choice when they laid out the PCB. The board schematic (http://www.keil.com) shows that port pins GPIOB 8 and 9 are used by the I2C interface. *Table 9* (Alternate Function Mapping) of the STM32F405xx and STM32F407xx Datasheets (DocID022152 Rev 6) shows that Port B Pins 8 and 9 are used by the AF2/3/4/5/9/11/13/15 alternate functions and AF5 connects instance I2C1. Similarly, the codec connections shown on the schematic and the Alternate Function Mapping (*Table 9*) mean we must use SPI2 as the I2S peripheral.

Information on sourcing the I2S clock can be found by referring to the clock tree in RM0090 Reference Manual (Doc ID 018909 Rev 6), *Figure 21*. If the I2S Phase Locked Loop (I2SPLL) is not running or an external I2S clock is not sourced, then we must enable the I2SPLL function, `I2S_Audio_Initialize()`:

```
RCC->CR |= RCC_CR_PLLI2SON;   /* Enable the PLLI2S */
               /* Wait till the main PLL is ready */
while((RCC->CR & RCC_CR_PLLI2SRDY) == 0) { }
```

As the SPI2 peripheral uses the APB1Periph clock, we also include the following:

```
__HAL_RCC_SPI2_CLK_ENABLE();
```

Configuring the GPIO pins and connecting the SPI2 AF is relatively straightforward; for example, we use `GPIO_Initialize()` as we did in earlier recipes. Note that we also need GPIO C Pin6 to source MCLK.

The final step is to initialize the I2S handle `struct` (defined in `stm32f4xx_hal_i2s.h`) with default values. A pointer to this structure is passed to the function named `HAL_I2S_Init()` that performs the low-level initialization. An important task within `HAL_I2S_Init()` is for the I2SPLL clock divider to give the desired I2S SCLK frequency.

The function used to initialize the codec named `CodecInitialize ( )` is very similar to the one that was presented in `codecDemo_c6v0` in the *Writing a driver for the audio codec* recipe in *Chapter 6, Multimedia Support*, but we've added some extra statements to allow this function to be used for either BEEP or SAMPLED audio. Similarly, `configureCodec()` also selects the appropriate setup.

The `main()` function super loop uses the function, `HAL_I2S_Transmit()`, to output audio samples representing a sinusoid. We can reuse the **Look-up-table** (**LUT**) that was introduced in `dacSinusoid_c5v0` from the *Generating a sine wave* recipe in *Chapter 5, Data Conversion* to represent the sampled sinusoid. However, as the I2S serial interface supports 16-bit signed samples, we'll need to convert the LUT to this format.

The I2S interface standard supports two (stereo) channels, and although we're operating the codec in mono (that is, channel A=B), we still need to transmit left and right samples, so each sample is repeated in the LUT array.

We've described the audio initialization in some detail and seem to have done a lot of work to produce very little so far, but judging from the number of posts on associated microcontroller internet forums, many novice embedded-system programmers have difficulty with this topic. Many developers use source code published by STM for their evaluation boards as a starting point, but they all tend to use different codec/microcontroller combinations, so reusing the code isn't always straightforward.

There's more...

Having generated a 'note', the question, *what frequency?*, arises. The I2S standard (Phillips Semiconductors, 1986) can help us answer this. The timing diagram depicted as follows illustrates an I2S data transmission:

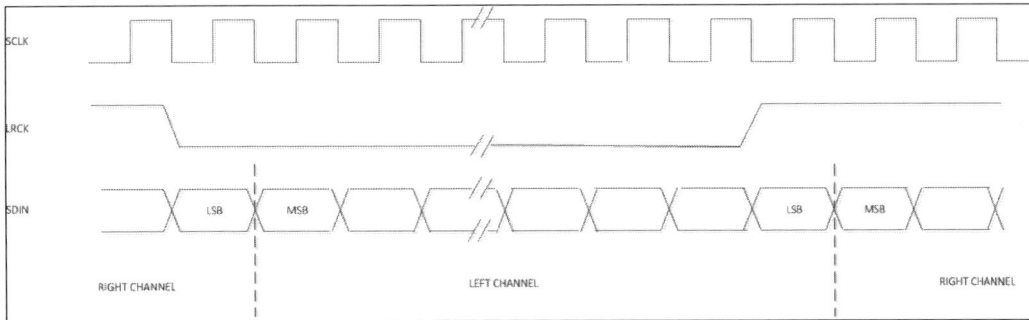

As the sinusoid is described by 20 samples and a sample frequency of 22 kHz (Fs), the period will be *20 × 10^(-3)⁄960.909 ms*, that is, a frequency of approximately 1.2 kHz. We can confirm this by connecting an oscilloscope to the audio jack.

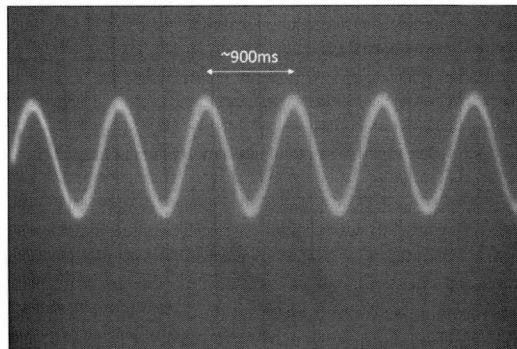

Currently, the main super loop only comprises one function call. We must be mindful that adding further statements in the loop may result in the I2S transmit register being starved.

How to play prerecorded audio

This recipe demonstrates how to play audio clips downloaded from the Internet globally. When you search for digital audio, you will encounter two common digital audio formats: **Waveform Audio File Format** (**WAVE** or **WAV**) and **MPEG-1, MPEG-2 Audio Layer III Format** (**MP3**). This recipe focuses on playing WAV-encoded audio clips. The STM3241G-EVAL and STM32F4-DISCOVERY evaluation boards both include an MP3 player demo that can be ported to other systems. This recipe illustrates a skeleton that could form the basis for a similar application on the MCBSTM32F400 evaluation board. We'll call this recipe `codecDemo_c7v1`.

Getting ready

The easiest way to import WAV audio samples into our program is to convert them into C source code (in the same way that images were imported in *Chapter 6, Multimedia Support*). A number of programs to manipulate WAV files and write samples to a C source file are available. This recipe uses a free converter by Colin Seymour called **WAVtoCode** that supports a number of WAV file formats. The following screenshot shows the conversion program being used (note that this program also includes a mixing desk):

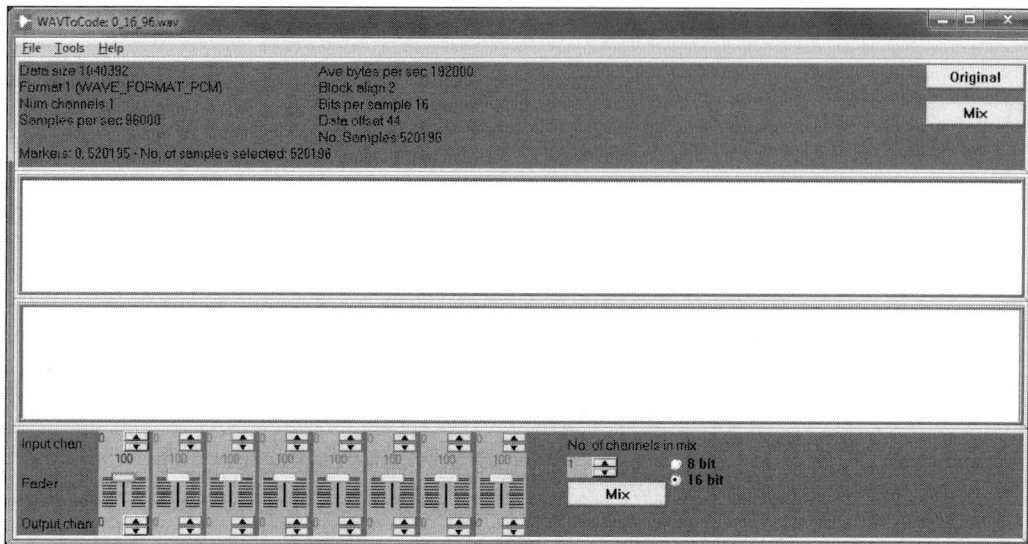

The program exports samples in 8/16 mono/stereo formats, as follows:

1. Download a 1-kHz WAV test signal sampled at 96 kHz (that is, *Fs = 96 kHz*) (http://www.rme-audio.com). Play the test signal using the converter, then select **16-bit Mix to Mono** from the **Tools** menu, and save as *signed 16-bit C Code*. A sample of the output is as follows:

```
BYTE data[NUM_ELEMENTS] = {
 -23417, -21874, -20238, -18517, -16716, -14844,
 -12909, -10920,  -8885,  -6811,  -4709,  -2586,
   -452,   1683,   3812,   5923,   8010,  10063,
  12073,  14033,  15931,  17763,  19519,  21193,
  22775,  24261,  25645,  26919,  28078,  29119,
  30037,  30825,  31483,  32008,  32397,  32648,
  32760,  32733,  32567,  32262,  31821,  31244,
  30535,  29696,  28730,  27642,  26438,  25121,
  23696,  22171,  20553,  18847,  17060,  15201,
  13279,  11299,...
```

More exciting audio clips are available!

2. Examine the output to confirm that the sinusoidal cycle repeats approximately every 96 samples (that is, approximately half a cycle is shown previously) giving a frequency of 1 kHz. Note: the size of the global array needed to store the samples exceeds the limit imposed by an unlicensed copy of uVision 5. *Chapter 9, Embedded Toolchain,* offers some open source compiler options that can be adopted to solve this problem.

How to do it...

Follow the outlined steps to play prerecorded audio:

1. Clone codecDemo_c7v0 from the *Configuring the audio codec* recipe that we described earlier in this chapter.

2. Store the test signal samples in a simple global array (note that the samples are duplicated for left and right channels), as follows:

```
int16_t data [] = {
    -23417, -23417, -21874, -21874, -20238, -20238,
    -18517, -18517, -16716, -16716, -14844, -14844,
    -12909, -12909, -10920, -10920, etc...
};
```

3. Open I2S_audio.c and change the sample frequency defined in the I2S_Audio_ Initialize() function to match that of the WAV file:

```
hi2s.Init.AudioFreq = I2S_AUDIOFREQ_96K;
```

4. Add a statement in I2S_Audio_Initialize() to enable interrupts:

```
NVIC_EnableIRQ(SPI2_IRQn);
```

5. Include the following **Interrupt Service Routine** (**ISR**) in the codecDemo.c file:

```
void SPI2_IRQHandler(void) {

    HAL_I2S_IRQHandler(&hi2s);
}
```

6. Include a transfer complete callback in the codecDemo.c file (that is, overriding this in stm32f4xx_hal_i2s.c):

```
void HAL_I2S_TxCpltCallback(I2S_HandleTypeDef *hi2s) {
    HAL_I2S_Transmit_IT(hi2s, (uint16_t *) dacLUT,
                                   ARR_SZ(dacLUT));

}
```

7. Modify the `main()` function so that it calls the `HAL_I2S_Transmit_IT()` function before entering the super loop (note that there is nothing left to do in the super loop as the interrupt service routine takes care of everything):

```
HAL_I2S_Transmit_IT(&hi2s, (uint16_t *) dacLUT,
                                ARR_SZ(dacLUT));

while (1) {
    if (mode == AUDIO_BEEP) {
      Beep(note);                    /* Play the note */
      wait_delay(500);                    /* pause */
    }
} /* WHILE */
```

8. Build, download, and run the program.

How it works...

The `HAL_I2S_Transmit()` function that we deployed in `codecDemo_c7v0` from the *Configuring the audio codec* recipe sends a block of audio samples to the codec. This function operates in polling mode to establish when the I2S transmit data register is empty, and it spins (busy waiting) on the codec's status register to determine when successive samples are needed. Unfortunately, while the processor is doing this, it can't perform much useful work. To address this problem, this recipe uses the `HAL_I2S_Transmit_IT()` library function to set the I2S interface to generate an interrupt when the I2S transmit data register is empty. It also keeps count of the number of samples that are transmitted and calls a function named `HAL_I2S_TxCpltCallback()` when the last audio sample in the block has been sent.

Prior to calling `HAL_I2S_TxCpltCallback()`, we need to enable interrupts (*step 4*), provide an interrupt service routine (*step 5*), and override the `HAL_I2S_TxCpltCallback()` function (*step 6*).

As the audio channel is essentially managed by the ISR, there isn't anything for the `main()` function to do!

Designing a low-pass digital filter

Joseph Fourier discovered that a complex signal could be described by a sum of sinusoids that is known as a Fourier series, and applying this idea enables us to visualize a signal frequency spectrum. A spectrum analyzer is a device that allows the frequency content of a signal to be displayed and measurements to be made. Two parameters, known as *magnitude* (amplitude) and *phase*, describe a *sinusoidal signal*. The magnitude spectrum describes the amplitude of each sinusoidal component that is summed, and the phase spectrum describes its associated phase. Often, we ignore the phase information and focus on the magnitude spectrum, but for some applications, particularly those that involve feedback, the phase of the signal is very important.

The magnitude spectrum of a *pure* 10 kHz sinusoidal signal is illustrated in the following diagram (the left panel) and that of a sampled version of the signal (the right panel):

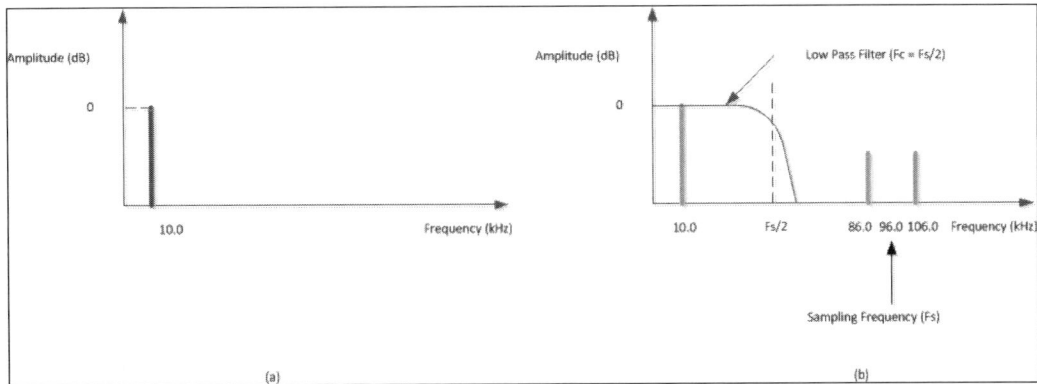

When we sample a signal, the steps in the digitized waveform (illustrated in *Chapter 5, Data Conversion*) introduce significant frequency components at higher frequencies. These appear as sidebands that are symmetrically displaced around integer multiples of the sampling frequency (Fs). As we saw in *Chapter 5, Data Conversion*, an analogue low-pass filter connected across the output of the D-A converter removes these harmonics and leaves the *pure* sinusoid.

The aim of digital filtering is to simulate the effect of analogue filters by writing a program that manipulates the digital signal samples. A digital filter is a function that accepts signal samples as inputs and returns samples that represent the processed signal in real time. In this case, real time implies that, if the input samples cannot be processed so as to produce output samples in a time frame 1/Fs, then the filter will fail.

We can only hope to provide an introduction to digital filters in this short chapter, and so we'll skip the preliminaries that are needed to gain a deeper understanding of this topic. Those motivated to find out more should consult an introductory text book.

Getting ready

The structure of a simple **Finite Impulse Response** (**FIR**) digital filter is shown next. It's called FIR because the output of the `y (n)` filter is only produced from input samples. FIR filters are inherently stable, but they cannot be implemented as efficiently as another class of digital filter, known as **Infinite Impulse Response** (**IIR**) filters. In IIR filters, the *y(n)* output is fed back and reused as another filter input. Potentially, this technique can produce instability, but this can be eliminated with careful design. We'll restrict ourselves to FIR designs here. In the following diagram, the block labeled T represents a time delay that is equal to the sample period, 1/Fs. So, in this case, the *y(n)* output is formed by the (equally-weighted) average of the current sample, *x(n)*, and four previous input samples: *x(n-1)*, *x(n-2)*, *x(n-3)*, and *x(n-4)*:

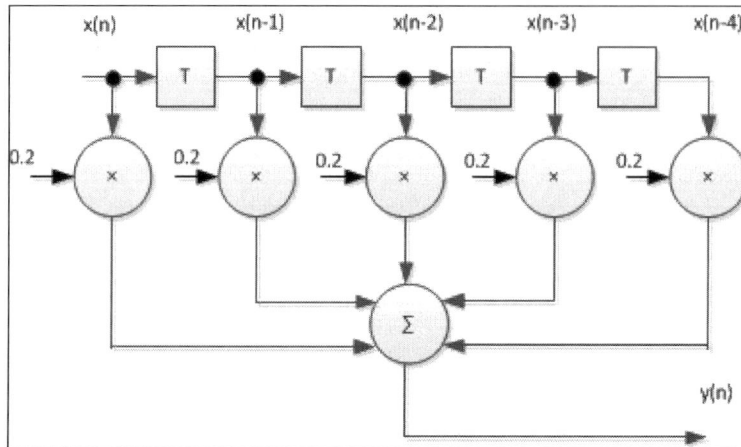

The output of a digital filter can be computed by a mathematical operation called *discrete convolution* and can be described mathematically, as follows:

$$y(n) = \sum_{k=0}^{n} x(n)h(n - k)$$

Here *x(n),y(n)* represent the input and output and *h(n-k) represents* the filter coefficients that are used to scale the input samples before they are summed. The number of coefficients used and their values determine the filter characteristic, and methods of calculating these parameters form the core of digital signal processing texts.

Rather than compute the filter weights longhand, which is rather tedious, we'll use a mathematical prototyping language called MATLAB to calculate them for us. Readers who do not have access to MATLAB could compute the filter coefficients using one of the techniques described in a digital signal processing text. Alternatively, there are a number of open source environments that are similar to MATLAB, such as GNU Octave, Sage, Scilab, and FreeMAT. The MATLAB script to design the filter is presented, as follows:

```
%%%%%%%%%%%%%%%%%%%%%%%%%%%%%%%%%%%%%%%%%%%%%%%%%%%%%%%%%%%%%%%%%%%
% MATLAB Script to generate low pass filter coefficients
%
% Mark.Fisher@uea.ac.uk
%%%%%%%%%%%%%%%%%%%%%%%%%%%%%%%%%%%%%%%%%%%%%%%%%%%%%%%%%%%%%%%%%%%
% set filter parameters
d= fdesign.lowpass('Fp,Fst,Ap,Ast',1000,2000,1,20,22000);
% design filter
Hd=design(d,'FIR');

% plot fiter respose
fvtool(Hd,'legend','on'); axis([0 22 -70 10])
```

The MATLAB script computes coefficients for a FIR filter having a pass band from < 1 kHz and a stop band from > 2 kHz. Attenuation in the pass band is < 1 dB, and in the stop band this is < 20 dB. The sampling frequency is 22 kHz. The filter's transfer function can be visualized using MATLAB's fvtool.

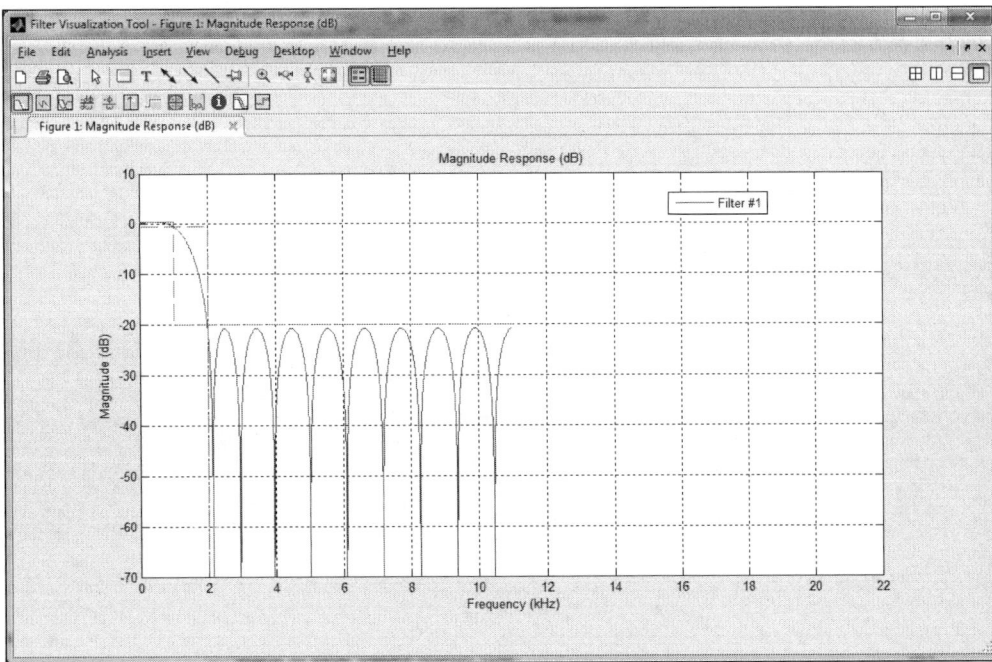

This frequency response confirms that our design meets the specifications. We can obtain the filter coefficients by plotting the filter impulse response, as shown in the next screenshot, and these can also be printed by the following MATLAB command prompt:

```
>> Hd.Numerator

ans =

Columns 1 through 9

-0.0537 -0.0138 -0.0071 0.0063 0.0259 0.0502 0.0767 0.1021 0.1233

Columns 10 through 18

0.1372 0.1421 0.1372 0.1233 0.1021 0.0767 0.0502 0.0259 0.0063

Columns 19 through 21

-0.0071 -0.0138 -0.0537
```

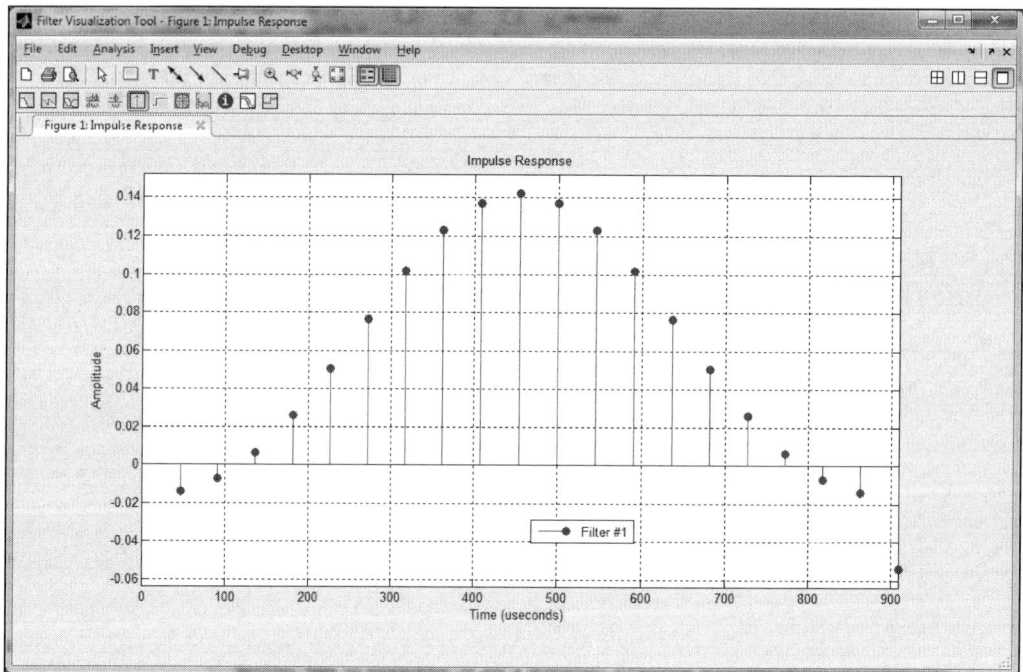

We'll implement this filter on our evaluation board as recipe `codecDemo_c7v2`.

How to do it...

1. Clone folder `codecDemo_c7v1` in the *How to play prerecorded audio* recipe. Change the RTE to include the LED (API).

2. Add the `filter()` function to the `codecDemo.c` file, as follows:

```
uint16_t filter(int16_t inSmpl) {
  /* Normalized Filter Coefficients */
  static const float lpfiltCoef[] =
     {    0.0, 0.0184, 0.0215, 0.0277, 0.0368, 0.0480,
       0.0602, 0.0720, 0.0818, 0.0882, 0.0905, 0.0882,
       0.0818, 0.0720, 0.0602, 0.0480, 0.0368, 0.0277,
       0.0215, 0.0184, 0.0 };
  static float smplBuff[] =
     { 0.0, 0.0, 0.0, 0.0, 0.0, 0.0, 0.0,
       0.0, 0.0, 0.0, 0.0, 0.0, 0.0, 0.0,
       0.0, 0.0, 0.0, 0.0, 0.0, 0.0, 0.0 };
  static uint8_t idx = 0;
  float lpVal = 0.0, outVal = 0.0;
  uint8_t coefIdx, newIdx;
  static const float int16max = (float) INT16_MAX;

  /* update buffer */
  newIdx = (nTaps-1-idx);
  smplBuff[newIdx] = (float) inSmpl;
  /* do convolution */
  for (coefIdx = 0; coefIdx<nTaps; coefIdx++)
    lpVal += lpfiltCoef[coefIdx] *
                    smplBuff[(newIdx+coefIdx)%nTaps];
  outVal = (int16_t)  (lpVal*sFactor * int16max);
  idx = (idx+1)%nTaps;

  return (uint16_t) outVal;
}
```

3. Change the ISR to write samples directly to the I2S data register:

```
void SPI2_IRQHandler(void) {

  if (flag) LED_On(0);
  else
    LED_Off(0);
  /* Transmit data */
  hi2s.Instance->DR = sample;
  flag = true;
}
```

4. Change the global LUT to hold audio data samples for a square wave (note: we're only storing data for one channel rather than a pair):

```
{ 0,   0,   0,   0, 0,
      0,   0,   0,   0, 0,
      1,   1,   1,   1, 1,
      1,   1,   1,   1, 1   };
```

5. Delete the `HAL_I2S_TxCpltCallback()` function.

6. Define global variables, as follows:

```
uint16_t sample = 0;
bool flag = false;
```

7. In `main()`, remove the call to `HAL_I2S_Transmit_IT()` and replace this with the following:

```
/* Enable I2S peripheral */
__HAL_I2S_ENABLE(&hi2s);

/* Enable I2S Interrupts */
__HAL_I2S_ENABLE_IT(&hi2s, (I2S_IT_TXE | I2S_IT_ERR));
```

8. Call `LED_Initialize()` in `main()` and remember to add `#include "Board_LED.h"`.

9. Add the code for the appropriate call to `filter()` within the super loop (note that, as the left and right channels carry the same signal, we only need to filter one):

```
while (1) {
  if (mode == AUDIO_BEEP) {
    Beep(note);                     /* Play the note */
    wait_delay(500);                        /* pause */
  }
  else {
    if (flag) {
      rightSmpl = i%2;
      if (!rightSmpl)               /* run filter */
        sample = filter(data[i>>1]);

      i++;
      i %= (sz<<1);                 /* MOD 2*sz */
      flag = false;
    }
  }
}
```

10. Build, download, and run the program.

How it works...

At the heart of the filter function is a mathematical operation known as convolution. This operation forms the sum of the current and previous 20 samples (that is, 21 in total), each of them is multiplied by a filter coefficient (weight). This is computationally demanding, and we're lucky that the ARM Cortex-M4 includes a floating point unit. This unit can perform single precision multiplications in three cycles (that is, ~18ns) plus the time needed for memory access. The most recent 21 input samples are stored in an array that is configured to operate as a circular buffer. A variable named `newIdx` identifies the oldest sample in the array and this sample is overwritten when a new sample becomes available. As it is critical that each sample is processed before it is written to the I2S transmit register, we clear a global boolean flag once the filter completes. If the ISR detects the flag set, then we switch an LED on to indicate an error. As time is critical, we chose to output samples directly to the I2S transmit register rather than use the `HAL_I2S_Transmit_IT()` library function. We chose to use a square wave as our test signal as it contains higher frequency harmonics. Note that the values of the filter coefficients (given by MATLAB) used in the program have been scaled, so they sum up to 1.0. We do this to avoid problems due to a possible overflow occurring when we assign the `outVal` variable. The following screenshot of an oscilloscope trace shows that the filter is recovering the fundamental frequency component (~1.2kHz) quite nicely with little evidence of distortion:

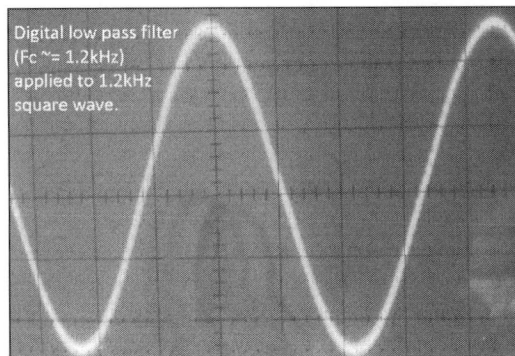

Digital low pass filter (Fc ~= 1.2kHz) applied to 1.2kHz square wave.

How to make an audio tone control

For the final recipe of this chapter, we'll make a digital tone control that emulates analogue circuits found on portable radios, and so on. Simple analogue tone circuits take the form of an active filter that uses a potentiometer to affect the filter transfer function (that is, emphasizing low/high frequencies—bass/treble—in the audio signal.

Although this recipe illustrates our filter operating in real time, it isn't the most efficient way of filtering digital audio. The audio codec includes its own DSP processing block, and this can be programmed to produce similar results more efficiently. We'll refer to this recipe as `codecDemo_c7v3`.

Getting ready

The high- and low-pass FIR filter coefficients that we need for this recipe are found using MATLAB. We've chosen the pass and stop bands that are shown in the following screenshot:

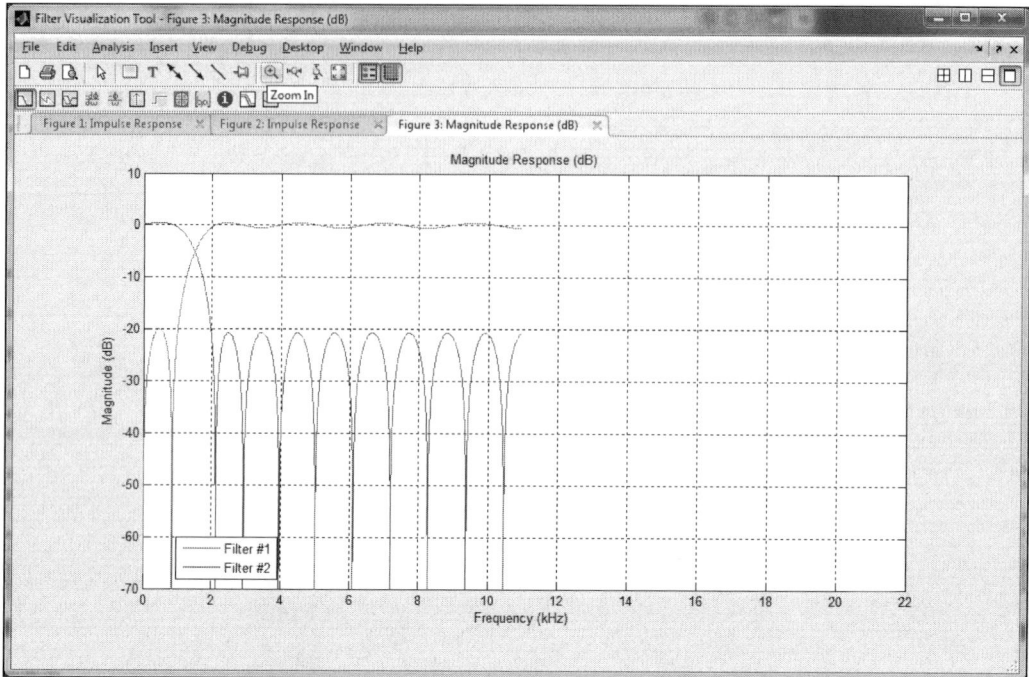

How to do it...

1. Clone codecDemo_c7v2 from the *Designing a low-pass digital filter* recipe and name the new folder codecDemo_c7v3.

2. Use the runtime management tool to add board support for the A/D converter. Add this statement to initialize the A/D converter:

    ```
    ADC_Initialize_and_Set_IRQ();
    ```

3. Add #include "Custom_ADC.h".

4. Include the Custom_ADC.c file in the project and copy the Custom_ADC.h file into the project folder. We developed these in adcISR_c5v0 from the *Setting up the ADC* recipe in *Chapter 5, Data Conversion*.

5. Add high-pass filter coefficients to the filter() function, as follows:

    ```
    static const float hpfiltCoef[] =
        { 0.0511, 0.0540, 0.0524, 0.0533, 0.0528, 0.0517,
          0.0493, 0.0450, 0.0363, 0.0   , 0.1000, 0.0637,
    ```

```
        0.0550, 0.0507, 0.0483, 0.0472, 0.0467, 0.0476,
        0.0460, 0.0489, 0.0 };
```

6. Modify the `filter()` function so that the output is formed by a weighted sum of low-pass and high-pass samples:

```
for (coefIdx = 0; coefIdx<nTaps; coefIdx++) {
    lpVal += lpfiltCoef[coefIdx] *
                    smplBuff[(newIdx+coefIdx)%nTaps];
    hpVal += hpfiltCoef[coefIdx] *
                    smplBuff[(newIdx+coefIdx)%nTaps];

}
outVal = (int16_t) ( (lpVal*sFactor +
                hpVal*((float)1.0-sFactor)) * int16max);
```

7. Add an ISR to service interrupts from the ADC, as follows:

```
void ADC_IRQHandler (void) {

    ADC3->SR &= ~2;          /* Clear EOC interrupt flag */
    adcValue = (ADC3->DR)>> 4; /* Get converted value */
    ADC3->CR2 |= (1 << 30);   /* Start next conversion */

}
```

8. Change the `main()` super loop so that we compute a global scale factor when we're not filtering the signal, that is, as follows:

```
if (flag) {
  rightSmpl = i%2;
  if (!rightSmpl)                     /* run filter */
    sample = filter(data[i>>1]);
  else                      /* update scalefactor */
    sFactor = ( (float) adcValue ) / c;

  i++;
  i %= (sz<<1);                 /* MOD 2*sz */
  flag = false;
}
```

9. Add the following global variables:

```
int32_t adcValue;
float sFactor = 0.0;
const float c = 255.0;
```

10. Build, download, and run the program.

How it works...

The output sample is a weighed sum of the low-pass and unfiltered signal. These weights depend on the ADC value that, in turn, reflects the position of the potentiometer thumbwheel. The computation of the scale factor ($0.0 \leq sFactor \leq 1.0$) involves division, and as this is more time-consuming than the multiply accumulate operation, we choose to do this when we're not running the filter.

There's more...

To implement convolution requires the multiplication and addition of real numbers. These operations are performed by the **Floating Point Unit** (**FPU**) of the Cortex-M4. Real numbers are represented using a floating-point binary format. Early computers used many different (manufacturer-specific) formats to represent real numbers, but nowadays formats are standardized. The IEEE 754-2008 standard defines two formats known as IEEE double- and single-precision. Our programs use the single-precision (32-bit) format by declaring variables of the `float` type. Numbers encoded using the double-precision format are declared using the `double` (64-bit) type. It is important to understand that the representations of floating-point numbers approximate the real values that they represent and the rounding errors introduced can be particularly problematic for DSP applications.

Early 16-bit microprocessors, such as Intel 8086, were unable to carry out arithmetic operations on floating point numbers without using a floating point library, and users who didn't purchase the additional 8087 coprocessor were faced with quite poor performance. However, in the last decade, integrated hardware FPUs have become more common. Convolutions, at the heart of DSP applications, make repeated use of **Multiply-Accumulate** (**MAC**) operations, and processors aimed at DSP applications, such as the Cortex-M4, include specific instructions that allow these to be executed very efficiently.

8
Real-Time Embedded Systems

In this chapter, we will cover the following topics:

- ▶ Multithreaded programs using event flags
- ▶ Multithreaded programs using mailboxes
- ▶ Why ensuring mutual exclusion is important when accessing shared resources
- ▶ Why we must use a mutex to access the GLCD
- ▶ How to write a multithreaded Pong game
- ▶ Debugging programs that use CMSIS-RTOS

Introduction

The title of the last chapter included the phrase, "*Real Time*". The term, Real Time, is used to describe a computing system that must meet deadlines. We did not define this term in *Chapter 7, Real-Time Signal Processing* because, in the context of handling audio samples, an implicit deadline is the sampling rate. However, you may recall that our ISR illuminated an error LED if the main super loop did not output the previous sample before a new sample arrived.

The audio application is an example of a soft deadline. It wouldn't be a catastrophe if the system missed this deadline once or twice; the audio quality would suffer, but this may go unnoticed. Contrast this with other applications, such as an embedded system used in fly-by-wire avionic applications, medical equipment, or a nuclear reactor. In these cases, missing a deadline could be catastrophic and result in death. Deadlines in these cases are known as hard deadlines and, in order to meet safety standards, designers need to guarantee that the system meets them. They may even be required to design redundancies to ensure that the system is robust to the catastrophic failure of a processor.

The last chapter illustrated that, although it is possible to design a simple real-time embedded system using a super loop, it gets increasingly tricky to ensure that deadlines are met as the system becomes more complex. An operating system is what is needed, but real-time systems do not use standard desktop operating systems, such as Windows or Linux, because it is impossible to guarantee that such systems will meet deadlines. Imagine a scenario where the pilot was landing an aircraft and the computer avionics system decided that now was a good time to defragment the hard disk! Instead of this, they use so-called **real-time operating systems** (**RTOS**), which are sometimes referred to as simply an embedded RTOS. Embedded RTOS are compact because the hardware running an embedded operating system is very limited in resources, such as RAM and ROM. Unlike a desktop operating system, the embedded operating system does not load and execute applications. This means that the system is only able to run a one application that is statically linked as a single executable image.

Operating systems based on the Linux kernel, and known as embedded Linux, are a popular choice as they are free from license fees. Embedded Linux forms the basis of the Android OS developed for smart phones and tablets. Many other examples of open source embedded RTOS exist. Most adopt the **Portable Operating System Interface** (**POSIX**) standard that supports open-standard application programming interfaces (APIs). We've adopted ARM's RTOS kernel, called RTX, as the RTOS used by examples in this chapter as it's included in the uVision5 IDE distribution. RTX was originally distributed as a **Real-Time Library** (**RL-ARM™**), designed to solve the real-time and communication challenges of embedded systems that are based on ARM processor-based microcontroller devices (refer to `www.keil.com/product/brochures/rl-arm_gs.pdf`). This library was recently revised and added to the CMSIS middleware standard and is now known as CMSIS-RTOS. A description of the API can be found at `https://www.keil.com/pack/doc/CMSIS/RTOS/html/index.html`, and advice on migrating from RL-ARM to CMSIS-RTOS is available here at `http://www.keil.com/appnotes/docs/apnt_264.asp`.

Support for multitasking is a key function of any operating system. Multitasking is a rudimentary form of parallel processing in which several tasks are run at the same time. Multitasking doesn't mean that tasks are executed in parallel. On a uniprocessor, there can be no true simultaneous execution of different tasks. Instead of this, the operating system switches between them, executing part of one task, then part of another, and so on. To the user, it appears that all the tasks are executing at the same time. The job of the operating system is to schedule each task on the CPU. The act of reassigning a CPU from one task to another one is called a context switch.

Embedded systems are at the heart of many everyday devices, such as smartphones, TVs, cameras, dishwashers, and so on. The system may comprise many tasks. For example, the dishwasher may embody a user interface, pump controller, and sensors. It is efficient to partition the software dealing with these elements into separate tasks. At any time, the total number of tasks can be divided into two groups: those that can be executed, and those suspended and waiting for an external event to occur (for example, the water temperature reaching a specific value or user input). RTOS supports preemptive scheduling, which allows tasks to be prioritized and can guarantee that some tasks in waiting will be given the CPU when an external event occurs.

We can illustrate how tasks are executed on the CPU by drawing an execution diagram. Consider three processes (just another name for tasks): a, b, and c that are executed periodically with a period of T seconds and have computation time (that is, they must be allocated on the CPU) of C seconds, as shown in the following table. The tasks are prioritized so that the task with the shortest period is allocated the highest priority (higher numbers imply a higher priority):

Process (P)	Period (T)	Computation Time (C)	Priority
a	50	12	1
b	40	10	2
c	30	10	3

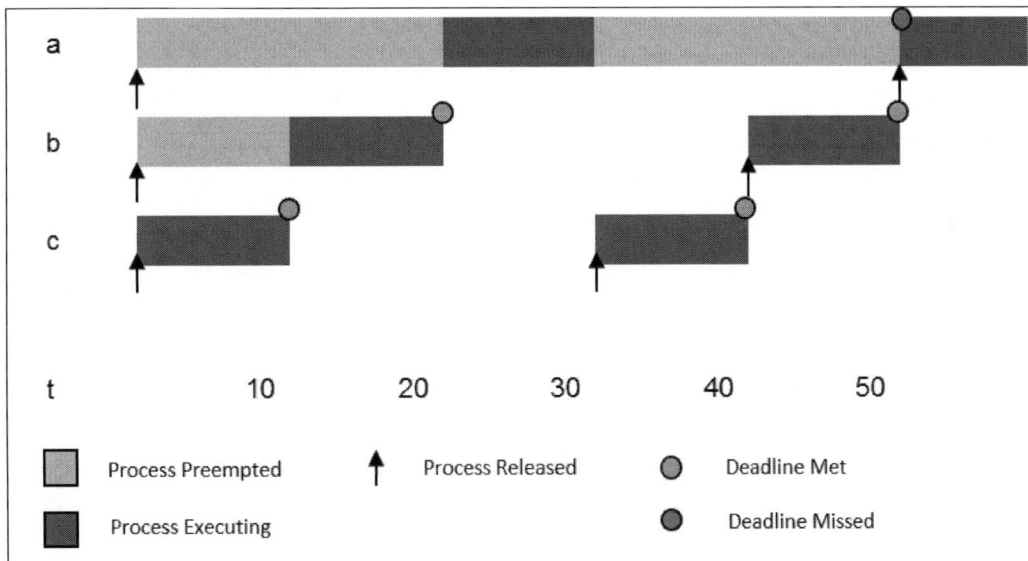

All the processes are released at t=0. Processes b and c are both meeting their respective deadlines. Process a gets preempted by b and c and misses its deadline. We can see from the execution trace that, in this case, the tasks cannot be successfully scheduled.

> We've assumed the context switch takes place instantly and the RTOS consumes no CPU time (in practice, both will incur some overhead).

Processes may need to share resources, and this raises the question of how they might communicate. The CMSIS-RTOS API solves both of these problems and more.

Multithreaded programs using event flags

This recipe will illustrate how to use CMSIS-RTOS to make an LED blink. We'll define two tasks or threads. The job of one task is to switch the LED ON, and the other one is to switch it OFF. The ON and OFF events are triggered by the tasks sending messages to each other. CMSIS-RTOS supports a number of intertask-communication strategies; our program uses event flags. We can illustrate our program using a state diagram, as follows:

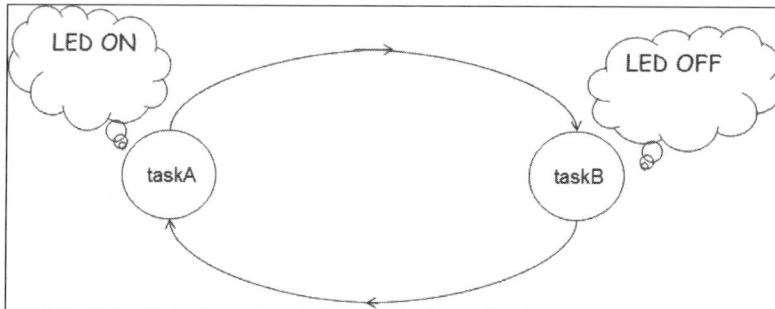

We'll call our first recipe, `RTOS_Blinky_c8v0`.

How to do it...

Create a new project (in a new folder) named `RTOS_Binky` and use the Run-Time Environment manager to select **Board Support → LED (API)** and **CMSIS → Keil RTX** as shown in the following screenshot. As usual, we can select **Resolve** to fix the warning messages. Note that this RTE is the same as the one that we introduced in *Chapter 2, C Language Programming*.

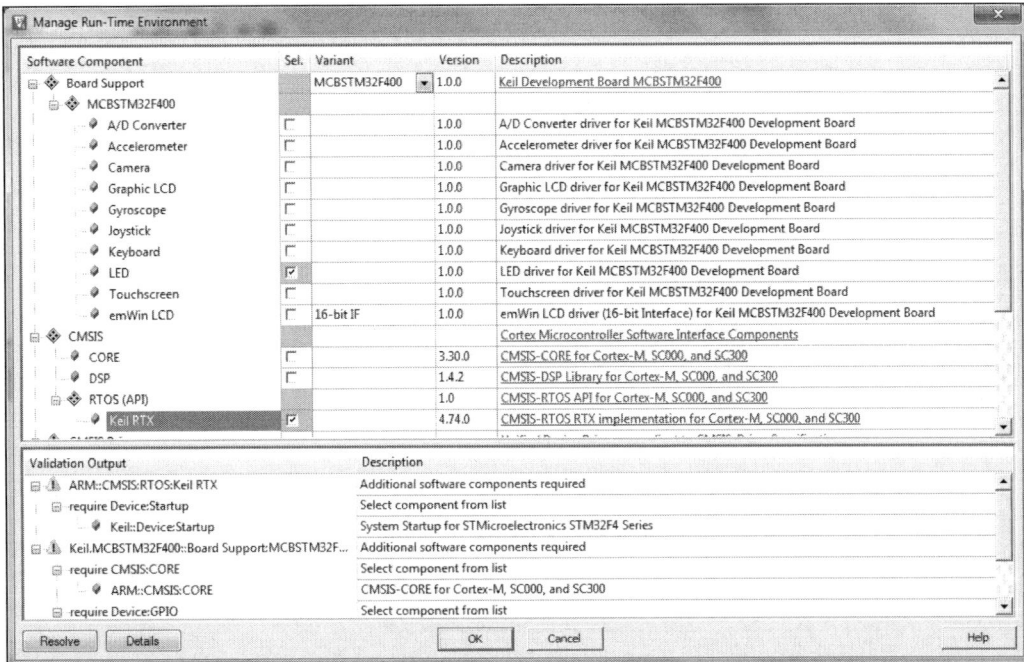

Software Component	Sel.	Variant	Version	Description
⊟ ◈ Board Support		MCBSTM32F400 ▼	1.0.0	Keil Development Board MCBSTM32F400
⊟ ◈ MCBSTM32F400				
◦ A/D Converter	☐		1.0.0	A/D Converter driver for Keil MCBSTM32F400 Development Board
◦ Accelerometer	☐		1.0.0	Accelerometer driver for Keil MCBSTM32F400 Development Board
◦ Camera	☐		1.0.0	Camera driver for Keil MCBSTM32F400 Development Board
◦ Graphic LCD	☐		1.0.0	Graphic LCD driver for Keil MCBSTM32F400 Development Board
◦ Gyroscope	☐		1.0.0	Gyroscope driver for Keil MCBSTM32F400 Development Board
◦ Joystick	☐		1.0.0	Joystick driver for Keil MCBSTM32F400 Development Board
◦ Keyboard	☐		1.0.0	Keyboard driver for Keil MCBSTM32F400 Development Board
◦ LED	☑		1.0.0	LED driver for Keil MCBSTM32F400 Development Board
◦ Touchscreen	☐		1.0.0	Touchscreen driver for Keil MCBSTM32F400 Development Board
◦ emWin LCD	☐	16-bit IF	1.0.0	emWin LCD driver (16-bit Interface) for Keil MCBSTM32F400 Development Board
⊟ ◈ CMSIS				Cortex Microcontroller Software Interface Components
◦ CORE	☐		3.30.0	CMSIS-CORE for Cortex-M, SC000, and SC300
◦ DSP	☐		1.4.2	CMSIS-DSP Library for Cortex-M, SC000, and SC300
⊟ ◈ RTOS (API)			1.0	CMSIS-RTOS API for Cortex-M, SC000, and SC300
◦ Keil RTX	☑		4.74.0	CMSIS-RTOS RTX implementation for Cortex-M, SC000, and SC300

Validation Output	Description
⊟ ⚠ ARM::CMSIS:RTOS:Keil RTX	Additional software components required
⊟ require Device:Startup	Select component from list
◦ Keil::Device:Startup	System Startup for STMicroelectronics STM32F4 Series
⊟ ⚠ Keil.MCBSTM32F400::Board Support:MCBSTM32F...	Additional software components required
⊟ require CMSIS:CORE	Select component from list
◦ ARM::CMSIS:CORE	CMSIS-CORE for Cortex-M, SC000, and SC300
⊟ require Device:GPIO	Select component from list

1. Create a new file named `RTXBlinky.c`, and create a skeleton by adding boilerplate code for `SystemClock_Config()`, and so on. Add this file to the project.

2. Select the **Configuration Wizard** tab for the `RTX_Conf_CM.c` file and configure the RTOS:

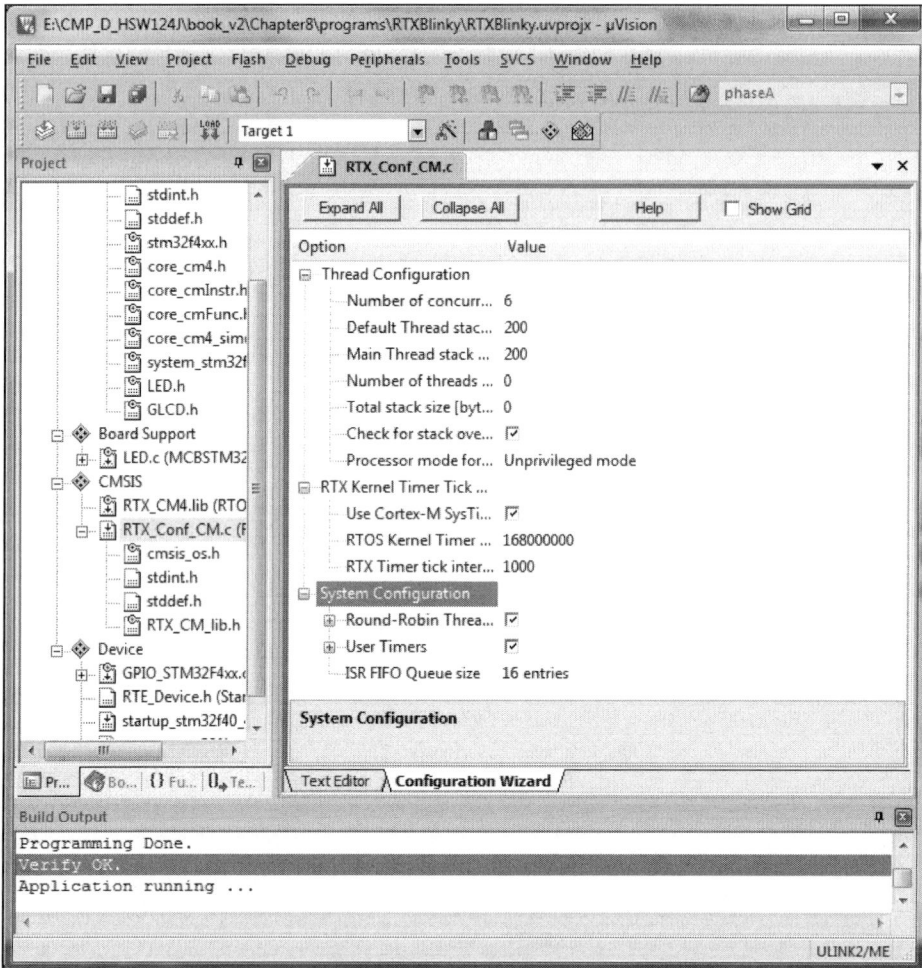

3. Open the `RTXBlinky.c` file and tasks A and B:

```c
#include "RTXBlinky.h"
/*-------------------------------------------------
 *      Thread 1 'taskA': Switch LED ON
 *-------------------------------------------------*/
void taskA (void const *argument) {
  for (;;) {
    /* wait for an event flag 0x0001 */
    osSignalWait(0x0001, osWaitForever);
```

```
      LED_On (LED_A);
      osDelay(500);
      /* set signal to taskB thread */
      osSignalSet(tid_taskB, 0x0001);
    }
}


/*-----------------------------------------------------
 *        Thread 2 'taskB': Switch LED OFF
 *----------------------------------------------------*/
void taskB (void const *argument) {
  for (;;) {
    /* wait for an event flag 0x0001 */
    osSignalWait(0x0001, osWaitForever);
    LED_Off (LED_A);
    osDelay(500);
    /* set signal to taskA thread */
    osSignalSet(tid_taskA, 0x0001);
  }
}


/*----------------------------------------------------
 *        Main: Initialize and start RTX Kernel
 *----------------------------------------------------*/
int main (void) {

  HAL_Init ();     /* Init Hardware Abstraction Layer */
  SystemClock_Config ();           /* Config Clocks */
  LED_Initialize();            /* Initialize the LEDs */

  tid_taskA = osThreadCreate(osThread(taskA), NULL);
  tid_taskB = osThreadCreate(osThread(taskB), NULL);

  /* set signal to taskA thread */
  osSignalSet(tid_taskA, 0x0001);

  osDelay(osWaitForever);
  while(1);
}
```

4. Create the `RTXBlinky.h` header file and add the following code:

```
#ifndef __RTX_BLINKY_H
#define __RTX_BLINKY_H

#include "stm32f4xx_hal.h"          /* STM32F4xx Defs */
#include "Board_LED.h"
#include "cmsis_os.h"

#define LED_A    0

/* Task ids */
osThreadId tid_taskA;
osThreadId tid_taskB;

/* Function Prototypes */
void taskA (void const *argument);
void taskB (void const *argument);

/* Define Threads */
osThreadDef(taskA, osPriorityNormal, 1, 0);
osThreadDef(taskB, osPriorityNormal, 1, 0);

#endif /* __RTX_BLINKY_H */
```

5. Build, download, and run the program.

How it works...

In RTOS, the basic unit of execution is a *task*. A task is very similar to a C procedure, but it must contain an endless loop:

```
void taskA (void const *argument) {
  for (;;) {
    /* taskA statements */
            }
}
```

So, a task never terminates and thus runs forever in a similar manner to the way that a program does. We can think of tasks as small self-contained programs. While each task runs in an endless loop, the task itself may be started by other tasks and stopped by itself or other tasks.

An RTOS-based program is made up of a number of tasks, which are controlled by the RTOS scheduler. The scheduler is essentially a timer interrupter that allots a certain amount of execution time to each task. So, task 1 may run for (say) 100 ms, then be descheduled to allow task 2 to run for a similar period of time; task 2 will give way to task 3; and finally, control passes back to task 1. If we open the **Configuration Wizard** tab for the `RTX_Conf_CM.c` file and expand the System Configuration menu, then we'll see that we're allocating slices of runtime to each task in a round-robin fashion, and tasks are switched every 5 ms (refer to the following screenshot):

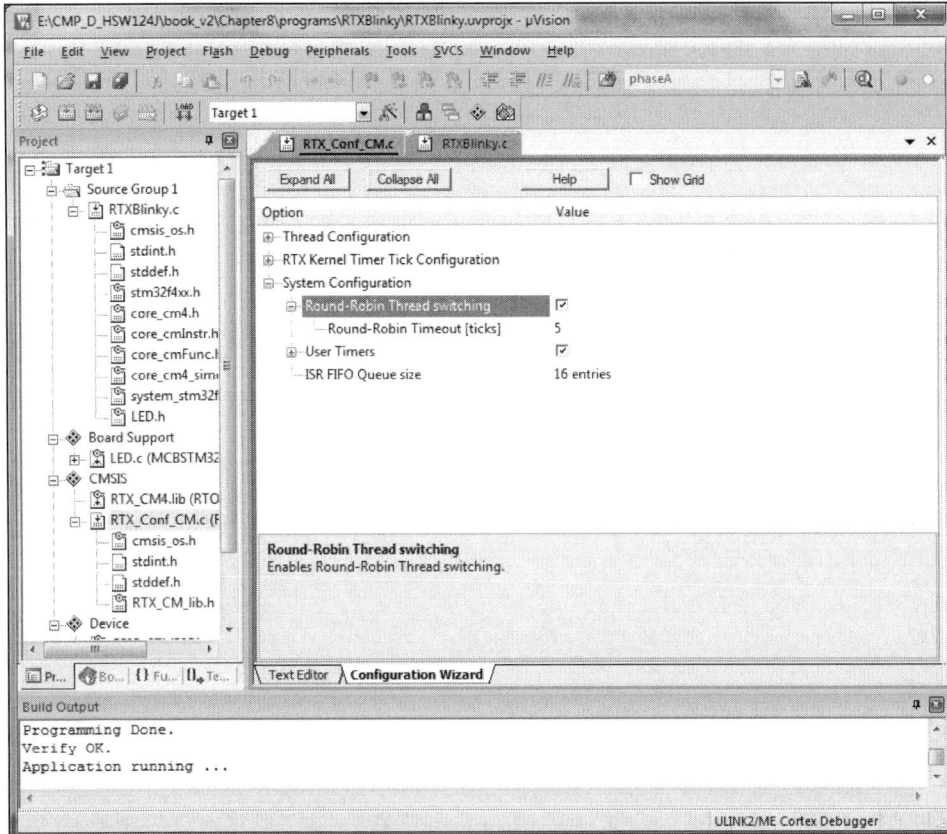

It is useful to think of all tasks running simultaneously, and each of them performing a specific function. This allows each functional block to be coded and tested in isolation and then integrated into a fully running program that, in turn, imposes structure and aids debugging. When a task is created, it is allocated its own task ID. This is a variable, which acts as a handle for each task and is used when we want to manage the activity of the task. We declare two such variables, one for `taskA` and one for `taskB`:

```
osThreadId tid_taskA;
osThread tid_taskB;
```

When CMSIS-RTOS runs on ARM-Cortex it uses the SysTick timer within the processor to provide the RTOS time reference. Each time we switch running tasks, the RTOS saves the state of all the task variables to a task stack and stores the runtime information about a task in a **Task Control Block** that is referenced by the task ID. In addition to the task variables, the Task Control Block also contains information about the status of a task. Part of this information is its run state.

A task can be in one of four basic states: **RUNNING, READY, WAITING,** or **INACTIVE.** Only one task can be running at a time, so the other tasks must be either READY, WAITING, or INACTIVE. A task is placed in the WAITING state when its execution is suspended. This may occur when it is waiting for an event to occur, such as a signal from another task. CMSIS-RTOS provides a number of mechanisms to enable tasks to communicate with each other, such as events, semaphores, and messages.

There may be many tasks that are READY for execution and it is the job of the scheduler to switch between them. CMSIS-RTOS is preemptive; the active thread with the highest priority becomes the RUNNING thread, provided that it is not waiting for any event. The initial priority of a thread is defined with the `osThreadDef()` function but may be changed during execution using the `osThreadSetPriority()` function. The function prototype for `osThreadSetPriority()` in the `cmsis_os.h` file identifies the function parameters, as follows:

```
/// \param     name       name of the thread fn.
/// \param     priority   initial priority of the thread fn.
/// \param     instances  number of possible thread instances.
/// \param     stacksz    stack size (bytes) for the thread fn.
```

Our program uses two threads, one to switch an LED ON and another to switch it OFF, so we define them, as follows:

```
osThreadDef(taskA, osPriorityNormal, 1, 0);
osThreadDef(taskB, osPriorityNormal, 1, 0);
```

> The `osPriorityNormal` argument is a pseudonym for the value, 0 (positive numbers indicate a higher priority, negative numbers a lower one).

Threads are created by the `osThreadCreate()` function, which returns a pointer to the Task Control Block. This function requires two arguments, a pointer to the thread definition and a pointer to its start argument. In our case, we write the following:

```
tid_taskA = osThreadCreate(osThread(taskA), NULL);
tid_taskB = osThreadCreate(osThread(taskB), NULL);
```

When each task is first created, it has sixteen event flags stored in the Task Control Block. It is possible to halt the execution of a task until a particular event flag or group of event flags are set by another task in the system. Our A and B tasks are very similar; the first statement in each is as follows:

```
osSignalWait(0x0001, osWaitForever);
```

This system call, suspends the execution of the task and places it into the WAIT_EVNT state. Any task can set the event flags of any other task in a system with the osSignalSet() CMSIS-RTOS function call. The main program statement is as follows:

```
osSignalSet(tid_taskA, 0x0001);
```

This statement sends a signal to taskA, which has been held by the following statement since this task was created:

```
osSignalWait(0x0001, osWaitForever);
```

The remaining taskA statements are as follows:

```
LED_on (LED_A);
osDelay(500);
osSignalSet(tid_taskB, 0x0001);
```

These statements turn the LED ON, invoke a delay, and then signal taskB. As well as running our application code as tasks, CMSIS-RTOS also provides some timing services, which can be accessed through CMSIS-RTOS function calls; osDelay() exemplifies the most basic of them. As CMSIS-RTOS ticks have been set at 1 ms, the delay is set at 0.5 seconds.

Multithreaded programs using mailboxes

The event flags that we saw in the last recipe can only been used to trigger the execution of tasks. In contrast to this, mailboxes support the exchange of program data between tasks. CMSIS-RTOS provides a mailbox system that buffers messages into mail slots and queues them between the sending and receiving tasks. This recipe, RTOS_Blinky_c8v1, provides an introduction to sending fixed-length messages between tasks using mailboxes.

How to do it...

1. Clone the RTOS_Blinky_c8v0 folder in the *Multithreaded programs using event flags* recipe that we described earlier.

2. Replace taskA( ) with the following function definition:

```
void taskA (void const *argument) {
  uint32_t i=0;
  for (;;) {
    mail_t *mail = (mail_t*)osMailAlloc(mail_box,
```

```
                                                    osWaitForever);
      mail->counter = i++;
      osMailPut(mail_box, mail);
      osDelay(1000);
    }
  }
```

3. Replace `taskB( )` with the following function definition:

```
void taskB (void const *argument) {
  for (;;) {
    osEvent evt = osMailGet(mail_box, osWaitForever);
    if (evt.status == osEventMail) {
      mail_t *mail = (mail_t*)evt.value.p;
      LED_Out(mail->counter);
      osMailFree(mail_box, mail);
    }
  }
}
```

4. Replace the `main()` function with the following:

```
int main (void) {

  HAL_Init ();     /* Init Hardware Abstraction Layer */
  SystemClock_Config ();          /* Config Clocks */
  LED_Initialize();           /* Initialize the LEDs */

  mail_box = osMailCreate(osMailQ(mail_box), NULL);

  tid_taskA = osThreadCreate(osThread(taskA), NULL);
  tid_taskB = osThreadCreate(osThread(taskB), NULL);

  osDelay(osWaitForever);
  while(1);
}
```

5. Declare the mailbox in the header file, `RTXBlinky.h`, by adding the following lines of code:

```
/* Mailbox */
typedef struct {
  uint32_t counter; /* A counter value              */
} mail_t;

osMailQDef(mail_box, 16, mail_t);
osMailQId  mail_box;
```

6. Build, download, and run the program.

How it works...

There are two tasks, named `taskA` and `taskB`. The role of `taskA` is to increment a counter, `taskB` displays the count value on the LEDs. The two tasks communicate by a mailbox, as shown in the following figure:

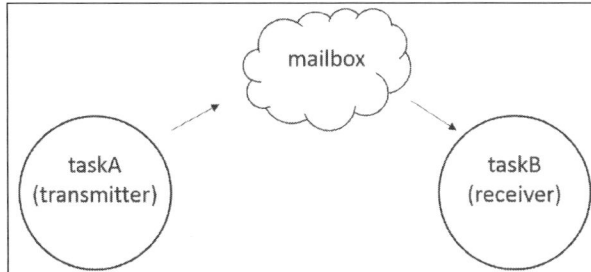

The message passed from `taskA` to `taskB` is declared as a struct named `mail_t`. The mailbox comprises a buffer that is formatted into a series of mail slots with pointers to each slot stored as an array. Take an example of the following statement:

```
osMailQDef(mail_box, 16, mail_t);
```

This statement creates a mail queue definition. We've chosen to use 16 mail slots, an arbitrary number that can be changed according to the complexity of our system. Sufficient memory is allocated to store 16 messages of type `mail_t`. Once defined, the following statement declares a mailbox variable:

```
osMailQId   mail_box;
```

The main function then creates and initializes the mail queue, assigning this variable:

```
mail_box = osMailCreate(osMailQ(mail_box), NULL);
```

The transmitter thread named `taskA()` calls `osMailAlloc( mail_box, osWaitForever)` to allocate a slot in the mailbox, and assigns a pointer to it. The second parameter represents a timeout value (we may need to wait for a slot to become free). The following statements assign a count value to the memory slot and put it in the mail queue:

```
mail->counter = i++;
osMailPut(mail_box, mail);
```

The receiver thread named `taskB( )` calls `osMailGet(mail_box, osWaitForever)` to check for messages in the mailbox. This function returns an event that contains mail information. Again, the second parameter represents a timeout (that is, there may be none). If there is a mail event, a pointer to the message data (that is, a `mail_t` struct) is assigned and the count is output to the LEDs. The following statement frees the memory slot:

```
osMailFree(mail_box, mail);
```

Further information on mailboxes can be found in the *CMSIS-RTOS API* (https://www.keil.com/).

Why ensuring mutual exclusion is important when accessing shared resources

A fundamental problem in multitasking is accessing shared resources. Text books often introduce this topic by considering the following problem. Imagine two tasks, both having access to a global variable. The job of one task, called an **incrementer**, is to increment the shared variable. The other task, called the **decrementer**, decrements the shared variable. The increment and decrement operations in each task are embedded within identical `for` loops. In this way, we arrange for the variable to be incremented and decremented the same number of times. The shared variable is reset to zero before the tasks are created and run. Once the tasks complete, one may expect the value of the shared variable to equal zero, as increment and decrement have been executed in equal measure by the two tasks. This recipe, named RTOS_Sem_c8v0, illustrates that, surprisingly, this is not the case.

How to do it...

1. Create a new project and using the manager configure the RTE to provide support for the Graphic LCD.

2. Add the following code to the project:

```
#include "RTXSem.h"

#define NCYCLES 500000        /* User Modified Value */
int sharedVar;                      /* Shared Variable */


/*------------------------------------------------------
 * Thread 1 'taskA': Increment Shared Variable
 *----------------------------------------------------*/
void taskA (void const *argument) {
  uint32_t p;
  bool flag = true;

  for (;;) {
    if (flag==true) {
      /* Inccrement the Shared Variable */
      for (p=0; p<NCYCLES; p++)
        sharedVar++;
      /* set signal to taskC thread    */
      osSignalSet(tid_taskC, 0x0001);
      flag = false;
```

```
      }
   }
}


/*--------------------------------------------------
 * Thread 2 'taskB': Decrement Shared Variable
 *------------------------------------------------*/
void taskB (void const *argument) {
  uint32_t p;
  bool flag = true;

  for (;;) {
    if (flag==true) {
      /* Decrement the Shared Variable */
      for (p=0; p < NCYCLES; p++)
        sharedVar--;
      /* set signal to taskC thread */
      osSignalSet(tid_taskC, 0x0002);
      flag = false;
    }
  }
}


/*--------------------------------------------------
 * Thread 3 'taskC': Display Shared Variable
 *------------------------------------------------*/
void taskC (void const *argument) {

  for (;;) {
    /* wait for an event flag 0x0003 */
    osSignalWait(0x0003, osWaitForever);
    GLCD_show_result(sharedVar);
    /* Kill Threads */
    osThreadTerminate (tid_taskA);
    osThreadTerminate (tid_taskB);
    osThreadTerminate (tid_taskC);
  }
}


/*--------------------------------------------------
 * Main: Initialize and start RTX Kernel
 *------------------------------------------------*/
int main (void) {
```

```
        HAL_Init ();    /* Init Hardware Abstraction Layer */
        SystemClock_Config ();              /* Config Clocks */

        GLCD_setup();

        sharedVar = 0;

        tid_taskA = osThreadCreate(osThread(taskA), NULL);
        tid_taskB = osThreadCreate(osThread(taskB), NULL);
        tid_taskC = osThreadCreate(osThread(taskC), NULL);

        osDelay(osWaitForever);
        while(1);
    }
```

3. Create file header file, `RTXSem.h`, and add the following code:

```
#ifndef __RTX_SEM_H
#define __RTX_SEM_H

#include "stm32f4xx.h"      /* STM32F4xx Definitions */
#include "RTXBlinkyUtils.h"
#include "cmsis_os.h"

/* Thread id of thread: task_a, b, c */
osThreadId tid_taskA;
osThreadId tid_taskB;
osThreadId tid_taskC;

/* Function Prototypes */
void taskA (void const *argument);
void taskB (void const *argument);
void taskC (void const *argument);

/* Thread Definitions */
osThreadDef(taskA, osPriorityNormal, __FI, 0);
osThreadDef(taskB, osPriorityNormal, __FI, 0);
osThreadDef(taskC, osPriorityNormal, __FI, 0);

#endif /* __RTX_SEM_H */
```

4. Create the `RTXBlinkyUtils.c` file, enter the following code, and add it to the project:

```
#include "RTXBlinkyUtils.h"

void GLCD_setup(void) {
```

```
    GLCD_Initialize();                    /* Initialise and */
    GLCD_SetBackgroundColor (GLCD_COLOR_WHITE);
    GLCD_ClearScreen ();                  /* clear the GLCD */
    GLCD_SetBackgroundColor(GLCD_COLOR_BLUE);
    GLCD_SetForegroundColor(GLCD_COLOR_WHITE);
    GLCD_SetFont (&GLCD_Font_16x24);
    GLCD_DrawString(0, 0*24, " CORTEX-M4 COOKBOOK ");
    GLCD_DrawString(0, 1*24, "  PACKT Publishing  ");
}

void GLCD_show_result(int value) {

    char buffer[128];

    GLCD_SetBackgroundColor(GLCD_COLOR_WHITE);
    GLCD_SetForegroundColor(GLCD_COLOR_BLACK);
    GLCD_DrawString (0, 3*24, "VAL =");
    sprintf (buffer, "%i   ", value);      /* make string */
    GLCD_DrawString (7*16, 3*24, buffer);  /* Display it */
}
```

5. Define the header file, `RTXBlinkyUtils.h`, and enter the following code:

```
#ifndef __RTX_BLINKY_GLCD_UTILS_H
#define __RTX_BLINKY_GLCD_UTILS_H

#include "Board_GLCD.h"
#include "GLCD_Config.h"
#include <stdio.h>
#include <stdlib.h>
#include <stdbool.h>

#define __FI    1                          /* Font index */

extern GLCD_FONT    GLCD_Font_16x24;

/* Function Prototypes */
void GLCD_setup(void);
void GLCD_show_result(int );

#endif /* __RTX_BLINKY_GLCD_UTILS_H */
```

6. Build, download and run the program.

[💡 Note the value of the shared variable output to the GLCD (it should be 0). Try running the program a few times.]

7. Change the value of NCYCLES, as follows:

    ```
    #define NCYCLES 500000
    ```

8. Build, download, and run the program. The value of the shared variable is output to the GLCD (it should be ≠ 0). Try running the program a few times.

How it works...

There are three tasks. Tasks A and B are incrementer and decrementer tasks, task C outputs the value of the shared variable to the GLCD. Task C waits for signals from both tasks, A and B, before calling the GLCD_show_result( ) function. To achieve this, task A sets flag 0x0001 and task B sets flag 0x0002; task C is released on flag 0x0003 (that is, the logical AND of task A and B flags).

To explain how the value of the shared variable can be anything other than zero, once the program terminates, we must consider how low-level machine instructions implementing increment or decrement operations will be executed for every possible scheduling of taskA and taskB. The increment operation involves reading a value from memory, storing it in one of the processor registers, adding one to it, and storing the result back in memory. Decrement will work in a similar way.

Assuming that the task switch between A and B always occurs after the task has written the updated value of the shared variable to memory, then the program operates successfully. When NCYCLES = 10, this will probably be the case. However, if the task switch occurs at the point just before the shared variable is written, then one task will be working with an outdated copy of the shared variable. This problem manifests as the error we observed.

There's more...

CMSIS-RTOS provides a solution to the problem of providing safe access to a shared resource (in this case a shared variable) by implementing a primitive known as a **Semaphore**. In general, a number of tasks (say, *p* tasks) may share a resource (that is, the resource can support a maximum of *p* tasks). To ensure that no more than p tasks access the resource at any time, we provide a variable (initialized to *p*) that will decrement each time a resource needs to use it and is incremented when the resource finishes with it. Thus, processes can only access the resource when *p>0*.

The case when a shared resource can only support one task (that is, *p=1*) can be managed by a binary semaphore called a **Mutual Exclusion** (**Mutex**). Mutexs are often used to ensure that critical sections of code are thread-safe. A piece of code is thread safe if it only manipulates shared data structures in a manner that guarantees safe execution by multiple threads at the same time. To ensure that the read, modify, or write operation produced by the increment or decrement is thread safe, we enclose the increment/decrement statement in task A or B as follows:

```
osMutexWait(mut_sharedVar, osWaitForever);
sharedVar++;
osMutexRelease(mut_sharedVar);
```

The variable named `mut_sharedVar` holds the semaphore. However, before we can use the semaphore, we must declare, register, and initialize it. The following recipe illustrates how this is done for a mutex used to control access to the GLCD. The same code statements can be used here; simply replace the `mut_GLCD` variable with `mut_sharedVar`. Once we've protected our critical section in this way, the program will run correctly and always return a value of zero, no matter how many cycles we specify.

Although the previous program is thread safe, there is another potential problem. Data is transmitted to the GLCD by a serial bus that is managed by functions that are defined in the GLCD library. If a task using the GLCD is switched while it is mid-way through writing to the GLCD, then there is a chance that the GLCD serial bus will stall and we'll lose data. This will manifest as a corruption of the screen and there is a chance that we'll misdiagnose this as a hardware fault, when in fact it is due to software. Many students try to fix this problem by arranging for all GLCD write statements to be in one task. This doesn't work because the serial bus is stalled as soon as a context switch occurs irrespective of what goes on in the other tasks. The solution is to treat the GLCD as a shared resource and enclose every invocation of the library code with calls to `osMutexWait( )` and `osMutexRelease( )`, even if they occur within the same thread. The following recipe illustrates this by emulating the `RTOS_Blinky_c8v0` folder in the *Multithreaded programs using event flags* recipe that we considered earlier in this chapter, this time using the GLCD to simulate the LEDs. We'll call this: `RTOS_Blinky_c8v2`.

Why we must use a mutex to access the GLCD

How to do it...

To access the GLCD using mutual exclusion, follow the steps outlined:

1. Create a new project and using the manager configure the RTE to provide support for the Graphic LCD.

2. Create a new file named `RTXBlinky.c`, add the boilerplate code, and then add this source file to the project.

3. Add the following code to `RTXBlinky.c`:

```c
#include "stm32f4xx_hal.h"         /* STM32F4xx Defs */
#include "RTXBlinkyUtils.h"
#include "cmsis_os.h"

osThreadId tid_taskA;       /* id of thread: task_a */
osThreadId tid_taskB;       /* id of thread: task_b */

osMutexId mut_GLCD; /* Mutex to control GLCD access */

/*-----------------------------------------------------
 *       Switch LED on
 *---------------------------------------------------*/
void switch_On (unsigned char led) {

  osMutexWait(mut_GLCD, osWaitForever);
  GLCD_SetBackgroundColor (GLCD_COLOR_WHITE);
  GLCD_SetForegroundColor(GLCD_COLOR_RED);
  GLCD_SetFont (&GLCD_Font_16x24);
  GLCD_DrawChar(led+(7*16), 4*24, 0x80+1);
  osMutexRelease(mut_GLCD);
}


/*-----------------------------------------------------
 *       Switch LED off
 *---------------------------------------------------*/
void switch_Off (unsigned char led) {

  osMutexWait(mut_GLCD, osWaitForever);
  GLCD_SetBackgroundColor (GLCD_COLOR_WHITE);
  GLCD_SetForegroundColor(GLCD_COLOR_RED);
  GLCD_SetFont (&GLCD_Font_16x24);
  GLCD_DrawChar(led+(7*16), 4*24, 0x80+0);
  osMutexRelease(mut_GLCD);
}


/*-----------------------------------------------------
 *       Thread 1 'taskA': Switch LED ON
 *---------------------------------------------------*/
void taskA (void const *argument) {
  for (;;) {
    osSignalWait(0x0001, osWaitForever);
    switch_On(LED_A);
    osDelay(500);
```

```
      osSignalSet(tid_taskB, 0x0001); /* signal taskB */
  }
}


/*--------------------------------------------------
 *      Thread 2 'taskB': Switch LED OFF
 *------------------------------------------------*/
void taskB (void const *argument) {
  for (;;) {
    osSignalWait(0x0001, osWaitForever);
    switch_Off(LED_A);
    osDelay(500);
    osSignalSet(tid_taskA, 0x0001); /* signal taskA */
  }
}


osMutexDef(mut_GLCD);

osThreadDef(taskA, osPriorityNormal, __FI, 0);
osThreadDef(taskB, osPriorityNormal, __FI, 0);


/*--------------------------------------------------
 *      Main: Initialize and start RTX Kernel
 *------------------------------------------------*/
int main (void) {

  HAL_Init ();    /* Init Hardware Abstraction Layer */
  SystemClock_Config ();          /* Config Clocks */

  GLCD_setup();

  mut_GLCD = osMutexCreate(osMutex(mut_GLCD));

  tid_taskA = osThreadCreate(osThread(taskA), NULL);
  tid_taskB = osThreadCreate(osThread(taskB), NULL);

  osSignalSet(tid_taskA, 0x0001);    /* signal taskA */

  osDelay(osWaitForever);
  while(1);
}
```

4. Create the `RTXBlinkyUtils.c` file, enter the following code, and add this to the project:

```c
#include "RTXBlinkyUtils.h"

void GLCD_setup(void) {

    unsigned char led;

    GLCD_Initialize();                      /* Initialize and */
    GLCD_SetBackgroundColor (GLCD_COLOR_WHITE);
    GLCD_ClearScreen ();                    /* clear the GLCD */
    GLCD_SetBackgroundColor(GLCD_COLOR_BLUE);
    GLCD_SetForegroundColor(GLCD_COLOR_WHITE);
    GLCD_SetFont (&GLCD_Font_16x24);
    GLCD_DrawString(0, 0*24, " CORTEX-M4 COOKBOOK ");
    GLCD_DrawString(0, 1*24, "  PACKT Publishing  ");
    GLCD_SetBackgroundColor(GLCD_COLOR_WHITE);
    GLCD_SetForegroundColor(GLCD_COLOR_RED);
    for (led=LED_A; led<LED_G+1; led++)
        GLCD_DrawChar((led+7)*16, 4*24, 0x80+0);
}
```

5. Modify `RTXBlinkyUtils.h` (defined in the previous recipe), accordingly.

6. Build, download, and run the program.

How it works...

Calls to GLCD functions within `switch_Off()` and `switch_On()` are protected by `mut_GLCD`, thus enforcing mutual exclusion. The `mut_GLCD` variable is declared as follows:

```c
osMutexId mut_GLCD;    /* Mutex to control GLCD access */
```

We also need to register the semaphore by including the following statement:

```c
osMutexDef(mut_GLCD);
```

We initialize this statement within `main()` by including the following:

```c
mut_sharedVar = osMutexCreate(osMutex(mut_GLCD));
```

How to write a multithreaded Pong game

To further illustrate how to use the features of CMSIS-RTOS that we've introduced in this chapter, we'll return to the Pong program that we first introduced in *Chapter 2, C Language Programming*. We'll call this recipe: `RTOS_Pong_v8v0`. Due to space limitations, we're only showing those parts of the code that are relevant to the RTOS implementation. Refer to *Chapter 2, C Language Programming* for details of helper functions defined in the `pong_utils.c` file.

How to do it...

To create a multithreaded pong game, follow the steps given:

1. Create a new project (new folder) called `RTOS_Pong`. Set the RTE to include board support for the ADC and GLCD. Include CMSIS-RTOS.

2. Create a file named `RTOS_Pong.c` and add a task to handle the GLCD:

```c
void taskGLCD (void const *argument) {
  BallInfo init_pstn = thisGame.ball;

  for (;;) {
    osEvent evt = osMailGet(mail_box, osWaitForever);
    if (evt.status == osEventMail) {
      mail_t *mail = (mail_t*)evt.value.p;
      thisGame.p1.y = mail->pdl;
      osMailFree(mail_box, mail);

      osMutexWait(mut_GLCD, osWaitForever);
      update_player();

      if (thisGame.ball.x<BAR_W) {    /* reset pstn */
        osDelay(T_LONG);
        erase_ball();
        thisGame.ball = init_pstn;
      }
      draw_ball();
      osMutexRelease(mut_GLCD);

      osDelay(T_SHORT);
      osSignalSet(tid_taskBall, 0x0001);
    }
  }
}
```

3. Add a task to update the ball and check for collisions:

```c
void taskBall (void const *argument) {

    for (;;) {
      osSignalWait(0x0001, osWaitForever);

      update_ball();
      check_collision();

      osSignalSet(tid_taskGLCD, 0x0001);
    }
}
```

4. Add a task to handle the ADC:

```c
void taskADC (void const *argument) {
uint32_t adcValue;
  for (;;) {
    mail_t *mail = (mail_t*)osMailAlloc(mail_box,
                                        osWaitForever);
    ADC_StartConversion();
    adcValue = ADC_GetValue ();

    mail->pdl = (adcValue >> 4) * (HEIGHT-BAR_H)/256;
    osMailPut(mail_box, mail);
    osDelay(T_SHORT);
  }
}
```

5. Add `main()`, save `RTOS_Pong.c`, and add the file to the project:

```c
int main (void) {
  HAL_Init ( );
  SystemClock_Config ( );

  game_Initialize();
  ADC_Initialize();
  GLCD_Initialize ();
  GLCD_Clear (White);             /* Clear the GLCD */
  GLCD_SetBackColor (White);   /* Set the Back Color */
  GLCD_SetTextColor (Blue);    /* Set the Text Color */

  mail_box = osMailCreate(osMailQ(mail_box), NULL);
  mut_GLCD = osMutexCreate(osMutex(mut_GLCD));

  tid_taskGLCD = osThreadCreate(osThread(taskGLCD), NULL);
```

```
    tid_taskBall = osThreadCreate(osThread(taskBall), NULL);
    tid_taskADC = osThreadCreate(osThread(taskADC), NULL);

    osDelay(osWaitForever);
    while(1)
        ;
}
```

6. Create an appropriate header file named `RTOS_Pong.h`:

```
#ifndef _RTOS_PONG_H
#define _RTOS_PONG_H

#include "cmsis_os.h"

#define __FI    1                   /* Font index 16x24 */

/* Mailbox */
typedef struct {
  uint32_t pdl; /* paddle position */
} mail_t;

osMailQDef(mail_box, 1, mail_t);
osMailQId  mail_box;

/* Mutex */
osMutexDef(mut_GLCD);
osMutexId mut_GLCD; /* Mutex to control GLCD access */

/* Function Prototypes for Tasks */
void taskGLCD (void const *argument);
void taskBall (void const *argument);
void taskADC (void const *argument);

/* Declare Task IDs */
osThreadId tid_taskGLCD;  /* id of thread: taskGLCD */
osThreadId tid_taskBall;  /* id of thread: taskGreq */
osThreadId tid_taskADC;   /* id of thread: taskMotor */

/* Define Threads */
osThreadDef(taskGLCD, osPriorityNormal, __FI, 0);
osThreadDef(taskBall, osPriorityNormal, __FI, 0);
osThreadDef(taskADC, osPriorityNormal, __FI, 0);

#endif /* _RTOS_PONG_H */
```

7. Copy the `pong_utils.c` and `pong_utils.h` files (refer to *Chapter 2, C Language Programming*.) and add these to the project.

8. Build, download, and run the program.

How it works...

The tasks named `taskGLCD( )` and `taskBall( )` are synchronized using a flag so the ball position is updated every time the screen is refreshed. The task named `taskADC( )` sends the position of the paddle to a mailbox; `taskGLCD( )` receives this value and uses it to render the paddle. The tasks are illustrated in the following diagram:

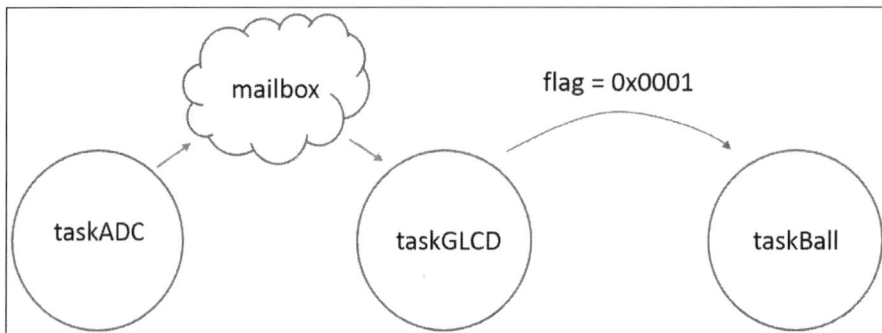

The tasks are loosely coupled and can be independently tested. For example, during debuging, the `taskADC( )` function and statements within `taskGLCD( )`, which read the mailbox and render the paddle, can be "commented out," leaving a simpler program that just moves the ball around the screen. The mailbox has only one slot. This is a key design decision that ensures that the paddle is rendered each time the ADC is read, so everything is synchronized to `taskADC()`.

Debugging programs that use CMSIS-RTOS

Using Keil's ULINK, we can gather and display general information about system resources while debugging our program.

How to do it...

1. Clone the `RTXBlinky` project that we described earlier in this chapter.

2. Select **Project → Options**. Under the **Debug** tab, select **Settings**.

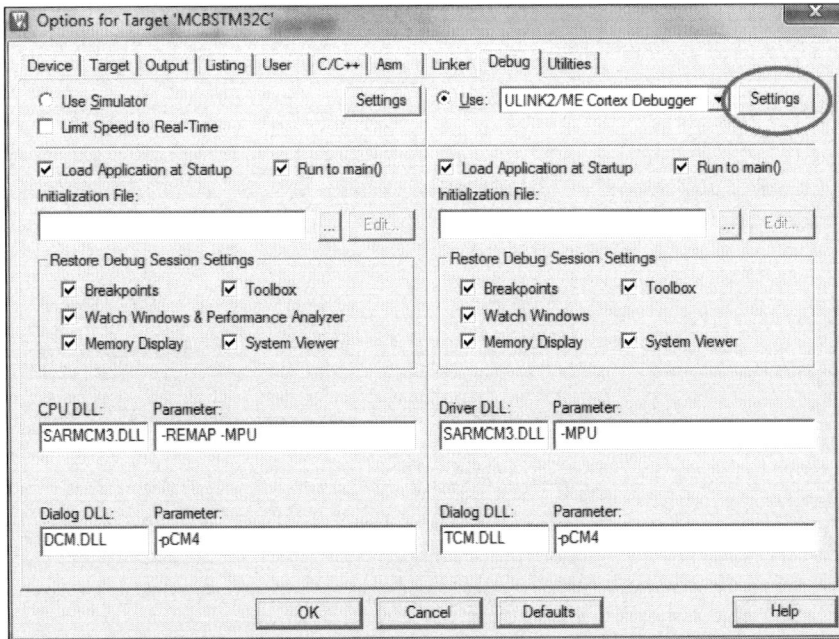

3. In the **Cortex-M Target Driver Setup** dialog, use the **Debug** tab to select the Serial Wire (**SW**) Communications protocol:

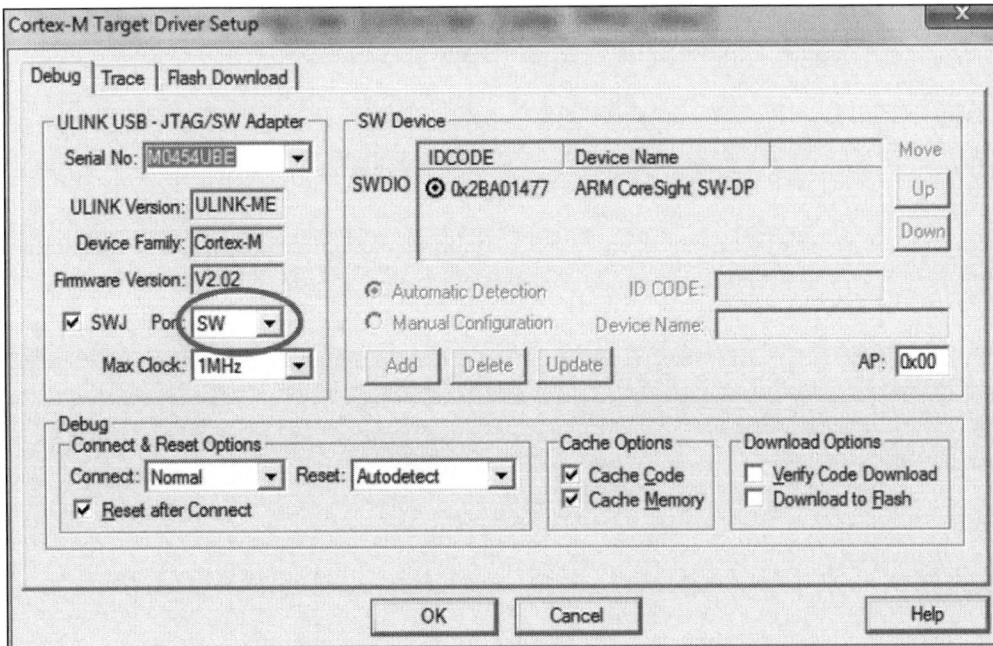

4. Still in the **Cortex-M Target Driver Setup** dialog, use the **Trace** tab to set the **Core Clock** frequency (**168.0 MHz**) and check **Trace Enable**:

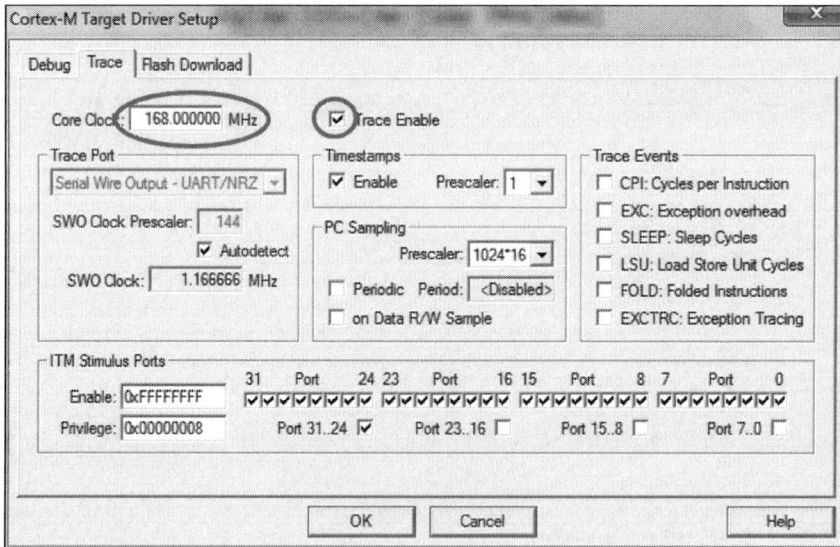

5. Download and run the program.

6. Debug the program by selecting **Debug → Start/Stop Debug Session** (*Ctrl+F5*).

7. Select **Debug → Run** (*F5*) to run the program.

8. Select **Debug → OS Support → System and Thread Viewer**.

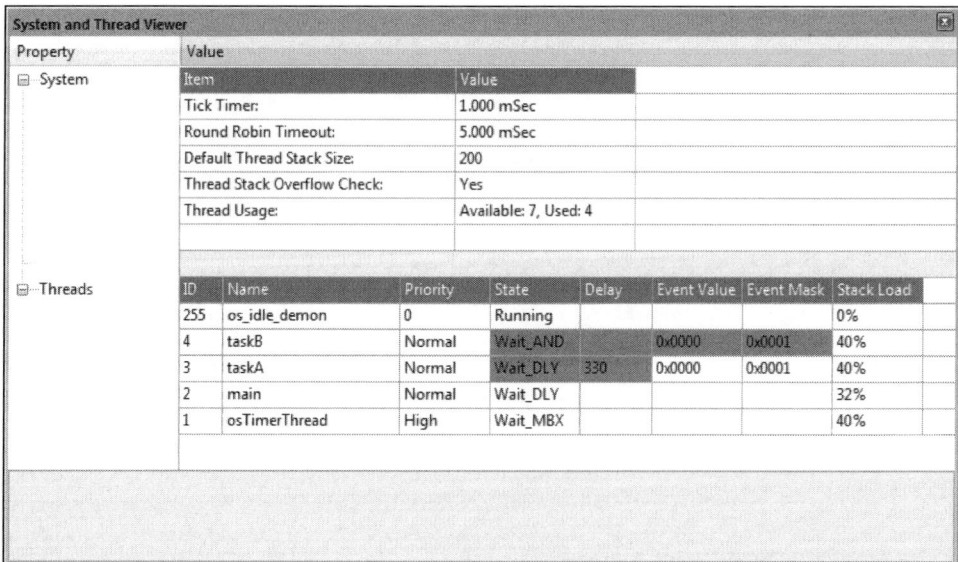

System and Thread Viewer

Property	Value
⊟ System	

Item	Value
Tick Timer:	1.000 mSec
Round Robin Timeout:	5.000 mSec
Default Thread Stack Size:	200
Thread Stack Overflow Check:	Yes
Thread Usage:	Available: 7, Used: 4

⊟ Threads

ID	Name	Priority	State	Delay	Event Value	Event Mask	Stack Load
255	os_idle_demon	0	Running				0%
4	taskB	Normal	Wait_AND		0x0000	0x0001	40%
3	taskA	Normal	Wait_DLY	330	0x0000	0x0001	40%
2	main	Normal	Wait_DLY				32%
1	osTimerThread	High	Wait_MBX				40%

> The cells that are highlighted in the previous screenshot are updated in real time as the program is running.

9. Select **Debug → OS Support → Event Viewer**. The cells that are highlighted in the following screenshot are updated in real time as the program is running:

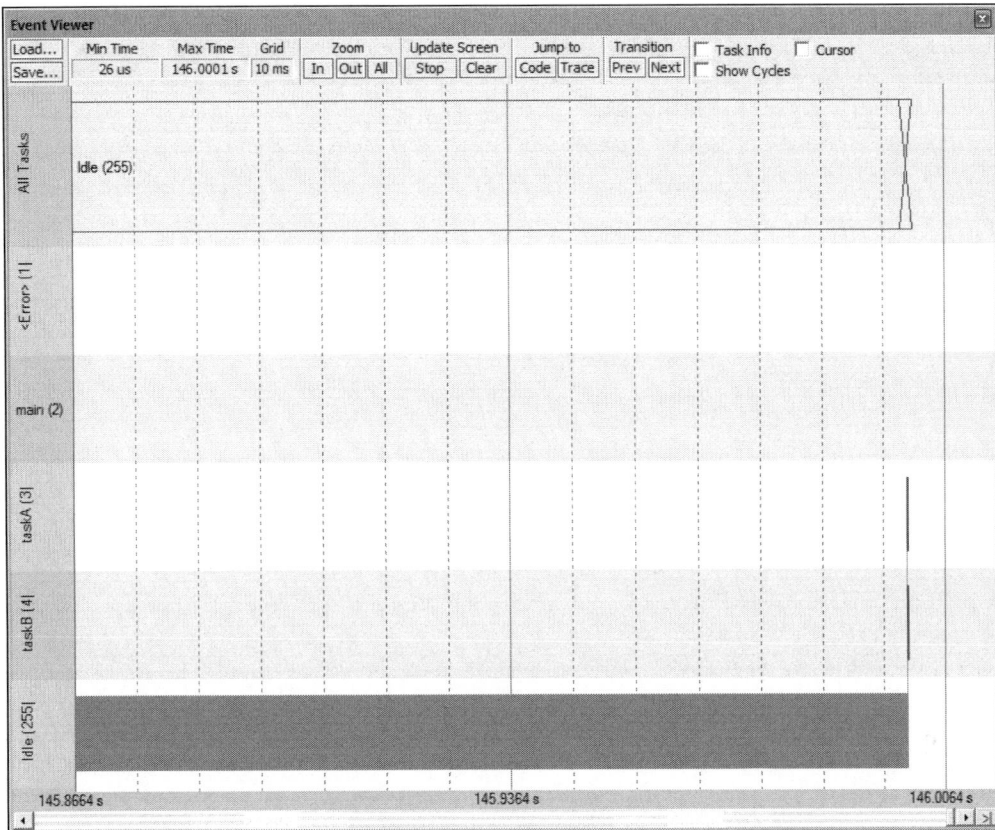

How it works...

The **System and Thread Viewer** window provides some useful information on System configuration and Threads. The values shown for the System reflect the ones that are defined in the `RTX_Conf_CM.c` file in the Configuration Wizard. There are a total of four threads, as CMSIS-RTOS manages `main()` and the `osTimerThread()` as discrete threads in their own right. When configuring the Trace (refer to *step 4*), it is very important to set the Core Clock frequency to agree with what is defined in `RTX_Conf_CM.c`:

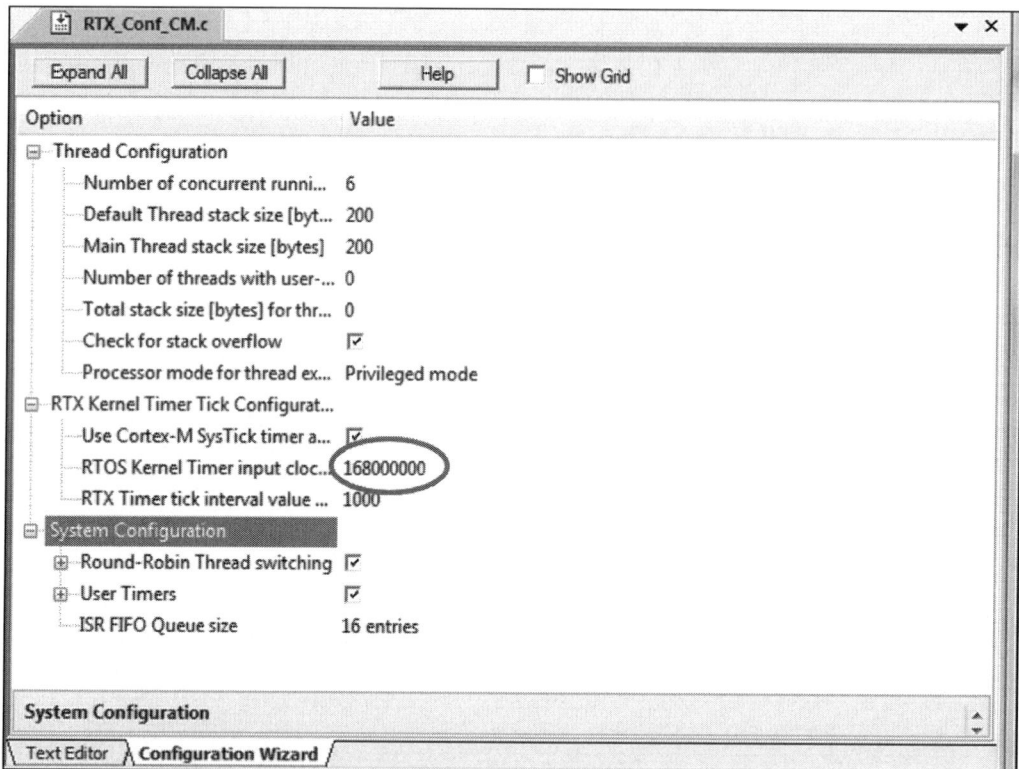

Further features of the debugger are discussed in *Keil Application Note No. 261* (refer to `http://www.keil.com/appnotes/files/apnt_261.pdf`).

9
Embedded Toolchain

In this chapter, we will cover the following topics:

- Installing GNU ARM Eclipse
- Programming the MCBSTM32F400 evaluation board
- How to use the STM32CubeMX Framework (API)
- How to port uVision projects to GNU ARM Eclipse

Introduction

A toolchain is a term that is used to describe a set of programming tools that are used to create a software product, which is typically an application program. A simple software development toolchain usually comprises a text editor, compiler, and linker, and often these are packaged together with other tools, such as a debugger, as an **Integrated Development Environment** (IDE). The ARM uVision5 IDE is very easy to use, but the constraints imposed on the free evaluation version and the relatively high cost of the licensed, professional version motivate many programmers to explore alternative, free, open source toolchains. Here is just a sample of the available alternatives:

- **emIDE**: This can be found at `http://www.emide.org/`
- **YAGARTO**: This can be found at `http://www.yagarto.org`
- **CooCox**: This can be found at `http://www.coocox.org/`
- **GNU ARM Eclipse**: This can be found at `http://gnuarmeclipse.github.io/`

Open source software is usually made available as source code and then released under a GNU General Public License. The GNU General Public License is intended to guarantee users the freedom to share and change all versions of a program, ensuring that it remains free software for all its users. Luckily, developers usually make precompiled versions of most software released under the GNU license available, often supporting the Windows, Linux, and Macintosh (OSX) operating systems.

However, installing and configuring an open source toolchain from a precompiled binary is not easy, so the aim of this chapter is to guide us through the process. We will illustrate the installation of the GNU ARM Eclipse toolchain on a Windows platform. We are choosing this route because the toolchain has recently migrated to GitHub and the installation guide has been revised.

Installing GNU ARM Eclipse

What is GNU ARM Eclipse? Well, Eclipse is an open source, integrated-development environment that can be configured for any toolchain. This is achieved, typically, by an extensible system of plug-ins that allows the environment to be customized. Eclipse is written mostly in Java, but plug-ins are available allowing it to be configured for a variety of languages. GNU ARM Eclipse plug-ins provide Eclipse CDT (C/C++ Development Tooling) extensions for GNU ARM toolchains, such as GNU Tools for ARM Embedded Processors, and others such as Linaro (`https://www.linaro.org/`), YAGARTO (`http://www.yagarto.org/`), and so on.

To install GNU ARM Eclipse, we need the following components:

- **The Eclipse IDE**: This is the IDE itself, and it can be found at `https://www.eclipse.org/`

- **GCC ARM Embedded Toolchain**: This is the GNU toolchain, and it an be found at `https://launchpad.net/gcc-arm-embedded`

- **Windows Build Tools**: These are the tools for make, rm, and so on (native to Linux), and they can be found at `https://github.com/gnuarmeclipse/windows-build-tools`

- **GNU ARM Eclipse plug-ins**: These are the plug-ins, and thy can be found at `https://github.com/gnuarmeclipse/plug-ins`

- **GNU ARM Eclipse QEMU Emulator plug-in**: This is an embedded processor emulator, and it can be found at `http://gnuarmeclipse.github.io/qemu/`

- **GNU ARM OpenOCD Debugging plug-in**: This is a debugging tool, and it can be found at `http://gnuarmeclipse.github.io/openocd/`

- **MDK-ARM Eclipse plug-in**: This is support for the U-Link debugger, and it can be found at `http://www.keil.com/support/man/docs/ecluv/default.htm`

Mostly, these are installed by downloading the latest version of their Windows installer
`.exe` file. As the MDK-ARM Eclipse plug-in only works with the Windows 32-bit version of
Eclipse, we chose 32-bit versions of the toolchain. The installation documentation provided is
comprehensive, so the following recipe (`GNU_ARM_Eclipse_Install_c9v0`) just gives us
an overview and links to the relevant web pages.

How to do it...

1. Follow the instructions at `http://gnuarmeclipse.github.io/toolchain/`
   `install/` and install the latest version (currently `gcc-arm-none-eabi-4_9-`
   `2015q3-20150921-win32.exe`) of the prebuilt GNU toolchain for ARM Embedded
   Processors. Execute the installer (in the final window, be sure to disable adding the
   toolchain path to the environment).

2. Test the gcc compiler by typing `"C:\Program Files (x86)\GNU Tools ARM`
   `Embedded\4.9 2015q3\bin\arm-none-eabi-gcc.exe" --version` in a
   command window:

```
Administrator: C:\WINDOWS\system32\cmd.exe

C:\Program Files (x86)\GNU Tools ARM Embedded\4.9 2015q3>"C:\Program Files (x86)
\GNU Tools ARM Embedded\4.9 2015q3\bin\arm-none-eabi-gcc.exe" --version
arm-none-eabi-gcc (GNU Tools for ARM Embedded Processors) 4.9.3 20150529 (re
lease) [ARM/embedded-4_9-branch revision 227977]
Copyright (C) 2014 Free Software Foundation, Inc.
This is free software; see the source for copying conditions.  There is NO
warranty; not even for MERCHANTABILITY or FITNESS FOR A PARTICULAR PURPOSE.

C:\Program Files (x86)\GNU Tools ARM Embedded\4.9 2015q3>
```

3. Refer to `http://gnuarmeclipse.github.io/windows-build-tools/`
   `download/`; download and run the latest version (currently `gnuarmeclipse-`
   `build-tools-win32-2.6-201507152002-setup.exe`) of Windows Build Tools
   from this link.

4. Check whether **Windows Build Tools** is functional by opening a command window in
   the folder where it was installed (that is, `"C:\Program Files\GNU ARM Eclipse\`
   `Build Tools\2.6-201507152002"`) and run `make --version` as follows:

```
C:\WINDOWS\system32\cmd.exe

C:\Program Files\GNU ARM Eclipse\Build Tools\2.6-201507152002\bin>make --version

GNU Make 4.1
Built for x86_64-w64-mingw32
Copyright (C) 1988-2014 Free Software Foundation, Inc.
License GPLv3+: GNU GPL version 3 or later <http://gnu.org/licenses/gpl.html>
This is free software: you are free to change and redistribute it.
There is NO WARRANTY, to the extent permitted by law.

C:\Program Files\GNU ARM Eclipse\Build Tools\2.6-201507152002\bin>_
```

5. Refer to `http://gnuarmeclipse.github.io/qemu/install/`, then download and run the latest version of the installer (currently `gnuarmeclipse-qemu-win32-2.3.50-201508041609-dev-setup.exe`) from this link.

6. Refer to `http://gnuarmeclipse.github.io/openocd/install/`, then download and run the latest version of the installer (currently `gnuarmeclipse-openocd-win32-0.9.0-201505190955-setup.exe`) from this link. Note that the documentation advises using the `SEGGER J-Link` debugger; other hardware is more difficult to set up.

7. Refer to `http://www.keil.com/support/man/docs/ecluv/default.htm` and install **MDK Version 5 - Legacy Support**.

8. Refer to `https://www.eclipse.org`, then download and run the latest version of the installer (currently Eclipse Mars. 1) from this link. Choose the version for C/C++ developers:

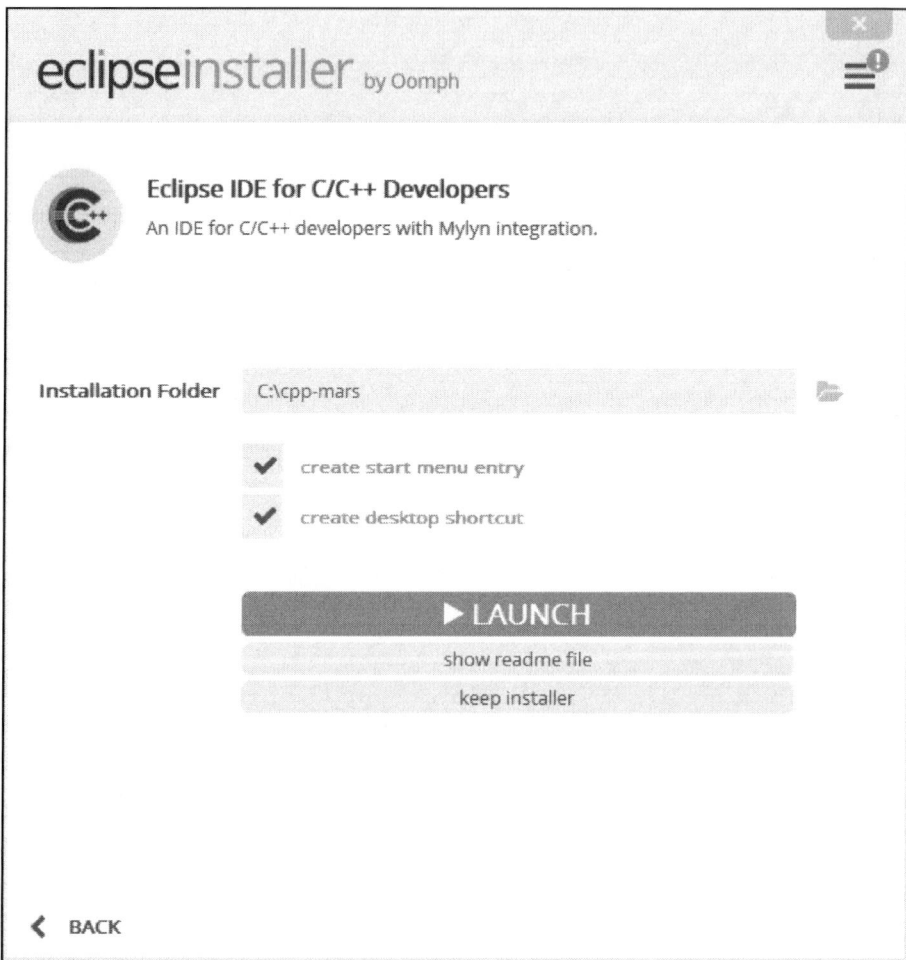

9. Refer to `http://gnuarmeclipse.github.io/eclipse/workspace/preferences/` and set the Eclipse preferences.

10. Refer to `http://gnuarmeclipse.github.io/plugins/install/` and install the GNU ARM Eclipse plug-ins using the standard Eclipse installer in the **Help →**
Install New Software menu. Note that, as we are working with Mars and we installed Eclipse configured for C/C++, then we may find that we already have some CDT tools (by default, plug-ins that are already installed are not displayed).

11. Refer to `http://gnuarmeclipse.github.io/plugins/packs-manager/`. To install packs, we need to select the pack perspective and find available packs, then install the ones that we want (make local copies). We're going to test our Eclipse IDE with the emulator configured as Discovery Board. So, we'll need the STM32F4 support pack:

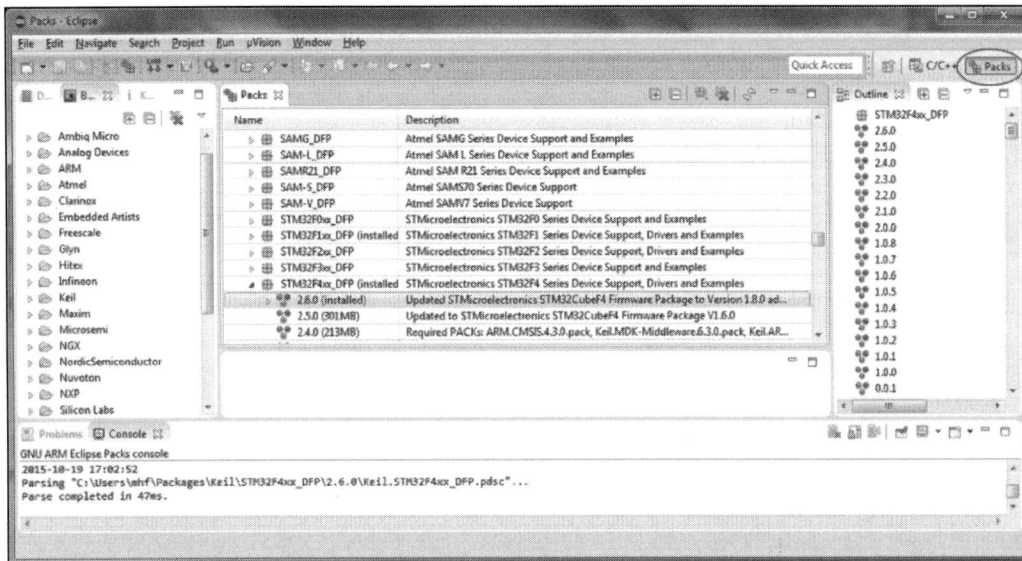

12. Refer to `http://gnuarmeclipse.github.io/tutorials/blinky-arm/` and use the wizard to create a Blinky ARM test project:

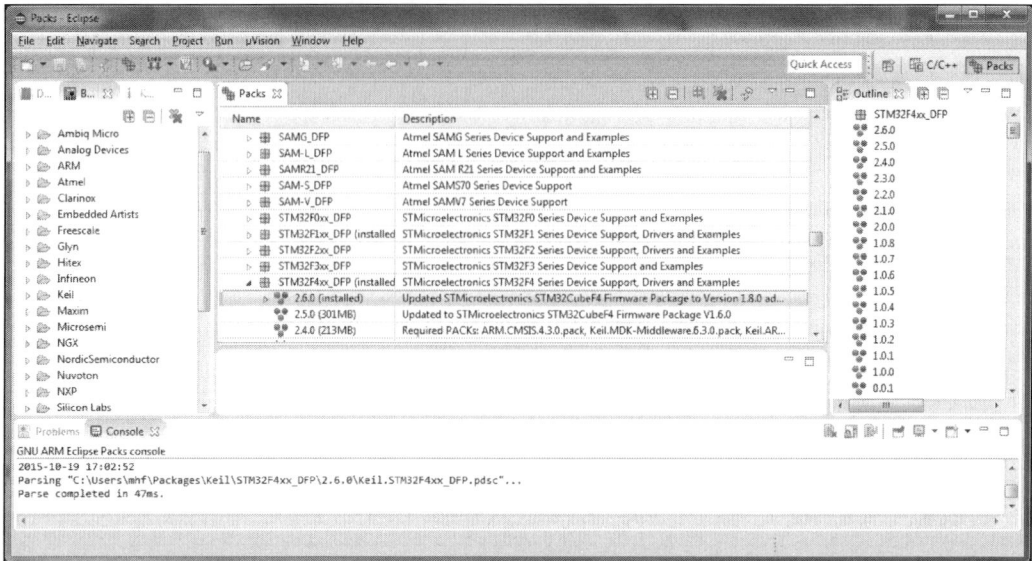

13. Refer to `http://gnuarmeclipse.github.io/tutorials/blinky-arm/`. Build the project and run the program on the **Discovery Board** emulator:

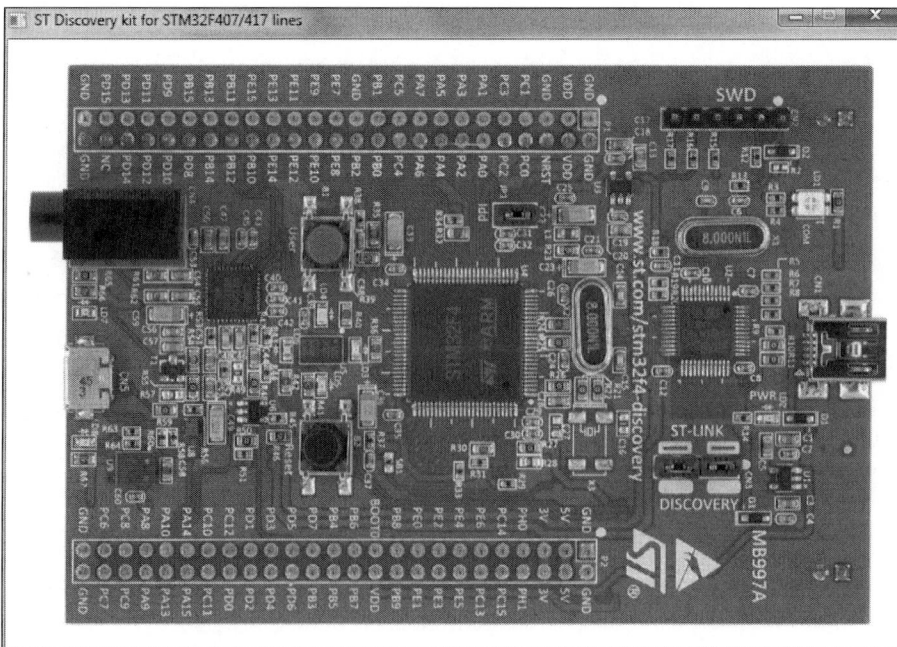

How it works...

Assuming that we successfully ran this code, then we have a working IDE. The Blinky wizard generates C++ code, so it may look a little strange. Don't worry; for the next recipe we'll create a C project.

Programming the MCBSTM32F400 evaluation board

This recipe will detail modifications that are necessary for the Blinky program created by the Eclipse project wizard and will show how to use the MDK-ARM Eclipse plug-in to flash the **STM32F407IG** part. We'll call this recipe GNU_ARM_Blinky_c9v0.

How to do it...

1. Invoke Eclipse.

2. The **MCBSTM32F400** evaluation board uses the STM32F407IG device, so we install the pack supporting this. To install the pack, switch to the **Packs** perspective and right-click the name of the pack:

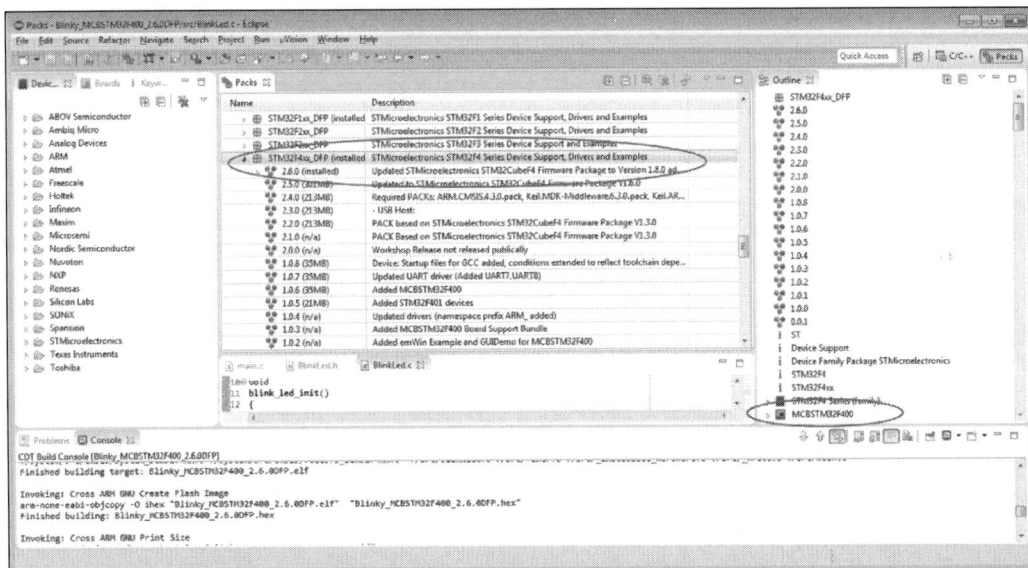

3. Refer to `http://www.keil.com/support/man/docs/ecluv/ecluv_flashSetup.htm` and install the MDK-ARM Eclipse plug-in. Note that, once this plug-in is successfully installed, the uVision icon and menu will appear in the toolbar:

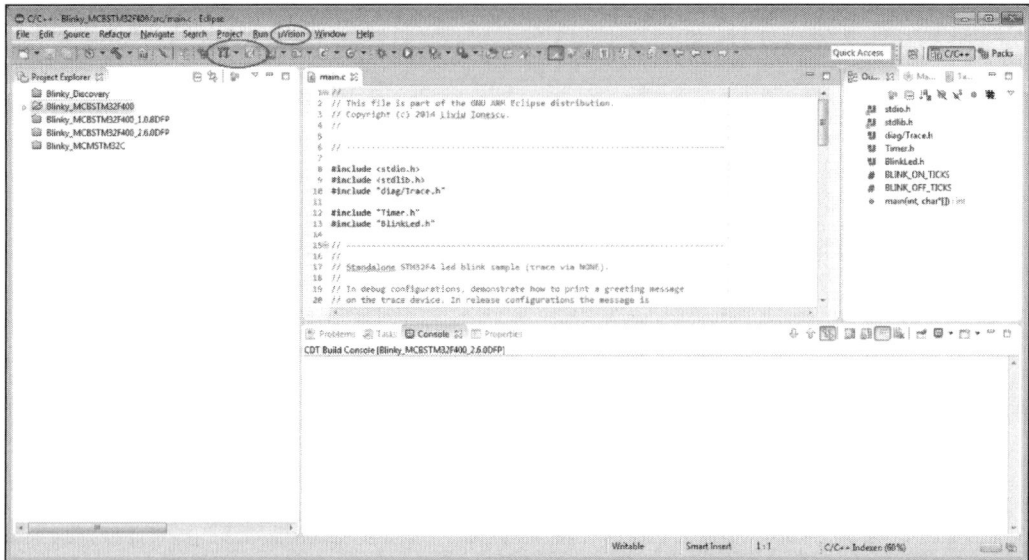

4. Switch to the C/C++ perspective. Select **File → New →C Project** and create a new project; give the project a name, select the **STM32F4xx** toolchain, and click **Next**:

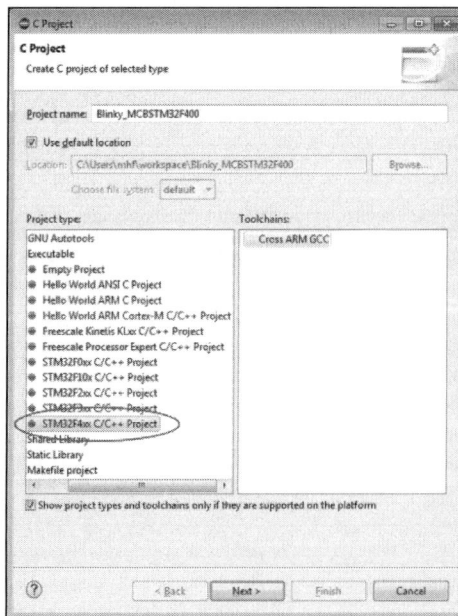

5. Choose the **STM32F407xx Chip Family**, and select **None (no trace output)** in **Trace output**:

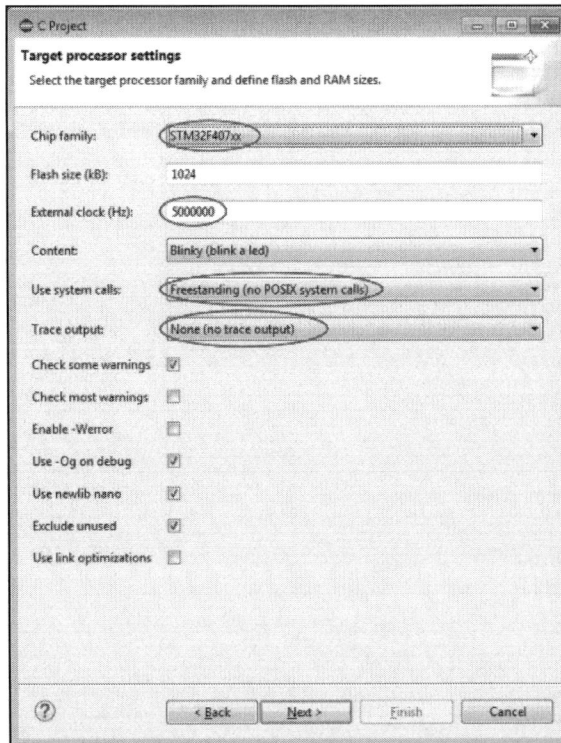

6. Open `BlinkLed.c`; in the `blink_led_int()` function, search for the following statement:

```
GPIO_InitStructure.Pull = GPIO_PULLUP;
```

Replace this statement with the following one:

```
GPIO_InitStructure.Pull = GPIO_PULLDOWN;
```

7. Open the header file named `BlinkLed.h`. Replace the `STM32F4DISCOVERY` definitions with the following:

```
// MCBSTM32F400 Eval. Board defs (led G6, active high)

#define BLINK_PORT_NUMBER        (6)
#define BLINK_PIN_NUMBER         (6)
#define BLINK_ACTIVE_LOW         (0)
```

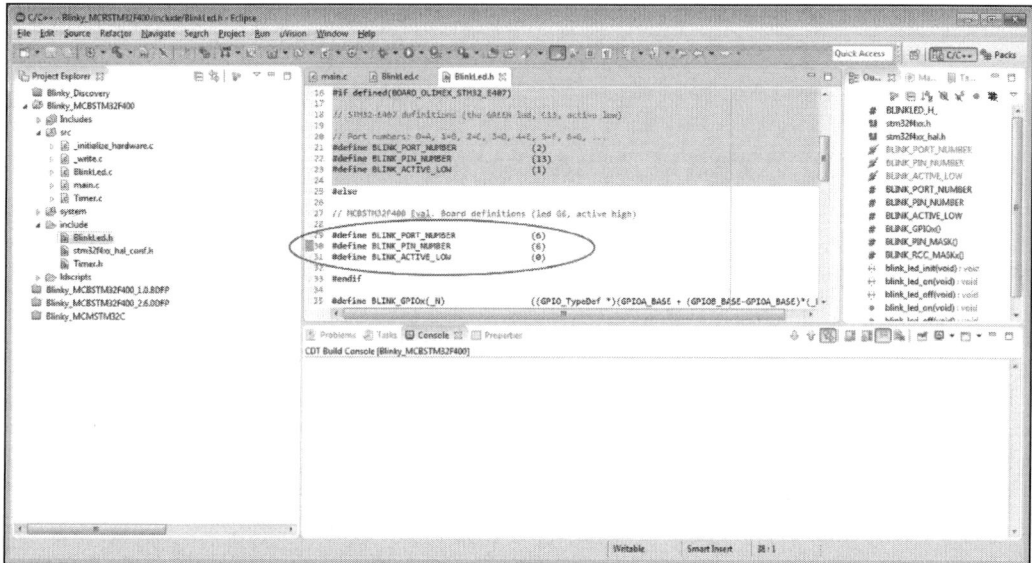

8. Select **Project** → **Build All** and build the project (or use the hammer icon shortcut).

9. Select **U-Link Load** → **Flash Download Configurations...** and create a new configuration as shown in the following screenshot. Note that selecting **Target Options** will open the familiar uVision project options dialog window.

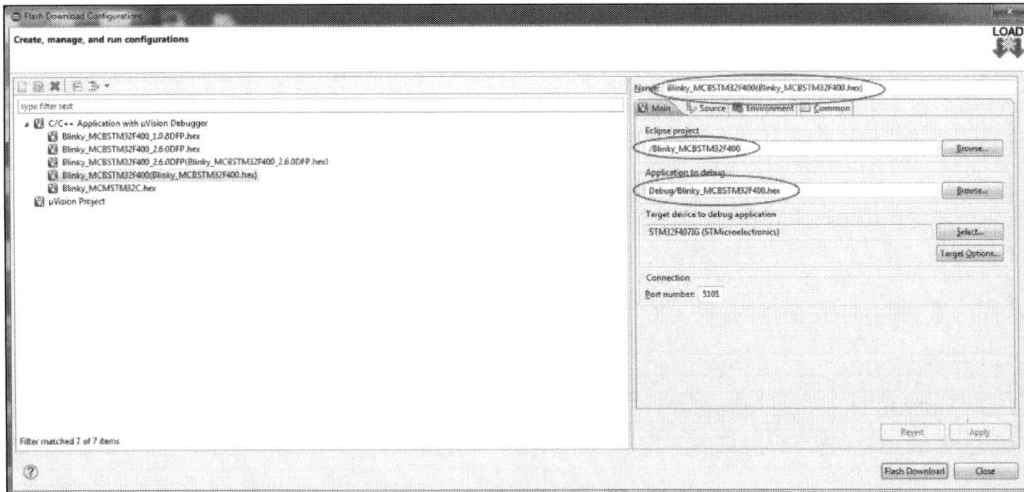

10. Select **Flash Download**. We may need to reset the board (depending on how we set the **Target Options**).

How it works...

We've simply configured the U-Link as a device programmer in this recipe. If you find that this doesn't work, then refer to http://www.keil.com/support/docs/3061.htm. Copy the .hex file created by Eclipse to a uVision project and use uVision to flash the board. You may need to use the UL2_EraseFW.exe utility that we discussed in *Chapter 2*, *C Language Programming*. If you do erase the U-Link firmware and subsequently flash the board using Eclipse, then expect the following to appear in the uVision **Output Console**:

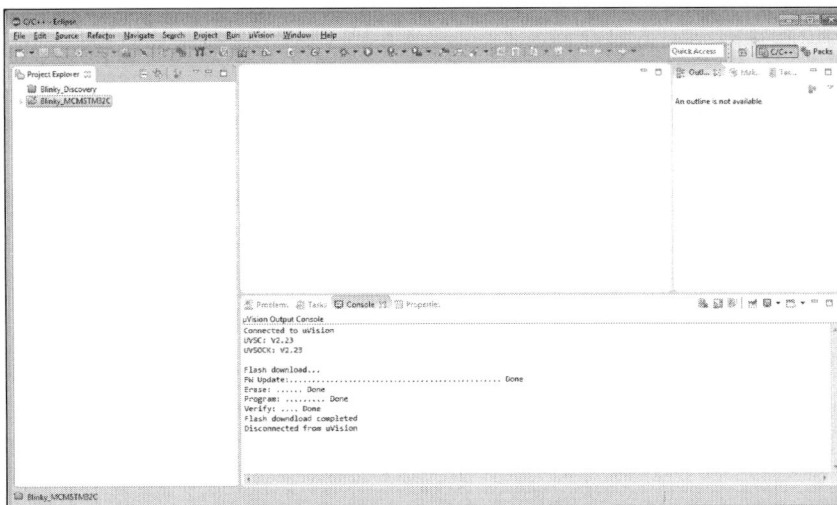

The calls to the `trace_printf()` function that appear in main can be ignored (or commented out). They are present to allow text strings to be displayed in the console debug window using a U-Link communication channel; however, although they work with the discovery board emulator, they don't with the U-Link2 hardware. This is not a serious problem because *Chapter 2, C Language Programming* describes other equally good approaches to debugging code.

You may have noticed that the GPIO support for LEDs provided by the Eclipse wizard is inferior to that in uVision. To drive multiple LEDs, we'll need to adapt some of the functions in the `LED.c` uVision file that is part of the `Hello_Blinky` project that we encountered in *Chapter 1, A Practical Introduction to Arm® Cortex®*.

How to use the STM32CubeMX Framework (API)

uVision5 provides two routes for users to configure their RTE. The first option, called Classic (used for all the recipes in Chapters 2-8), configures the STM's **Hardware Abstraction Layer** (**HAL**) using the `RTE_Device.h` header file. This option allows users to quickly configure the RTE for most CMSIS-enabled devices. The second option uses STM's graphical configuration tool, STM32Cube MX, to perform low-level configuration of the HAL directly. Example projects using both approaches are shipped with recent versions of Device Family Packs (for example, DFP 2.6.0). This recipe (named `ARM_STM32CubeMX_Blinky_c9v0`) shows you how to build a Blinky project using STM's tool.

How to do it...

1. Create a new project named `STM32CubeMX_Blinky`. Choose the **STM32F407IGHx** device.

2. Configure the RTE for the MCBSTM32F400 board. Check the **Board Support → LED (API) and Device → STM32Cube Framework (API) → STM32CubeMX** options. Then, select **Resolve** and **OK**.

3. If you haven't installed **STM32CubeMX** yet, you will be prompted to do so. It is freely available from `www.st.com` (search for *STM32Cube initialization code generator*).

4. If you have installed STM32CubeMX, then you should see this window asking you to launch the program:

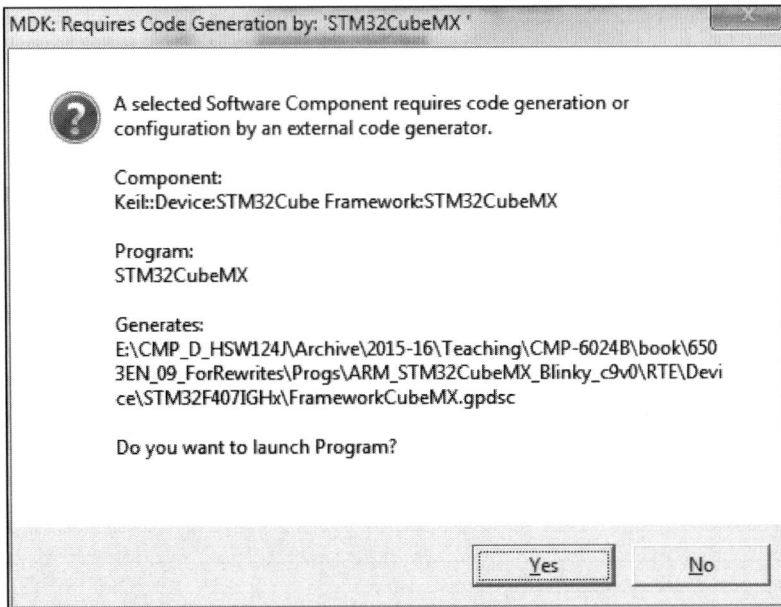

MDK: Requires Code Generation by: 'STM32CubeMX'

A selected Software Component requires code generation or configuration by an external code generator.

Component:
Keil::Device:STM32Cube Framework:STM32CubeMX

Program:
STM32CubeMX

Generates:
E:\CMP_D_HSW124J\Archive\2015-16\Teaching\CMP-6024B\book\650
3EN_09_ForRewrites\Progs\ARM_STM32CubeMX_Blinky_c9v0\RTE\Devi
ce\STM32F407IGHx\FrameworkCubeMX.gpdsc

Do you want to launch Program?

[Yes]   [No]

5. Once **STM32CubeMX** is launched, you should see the initial welcome screen. Choose **New Project**:

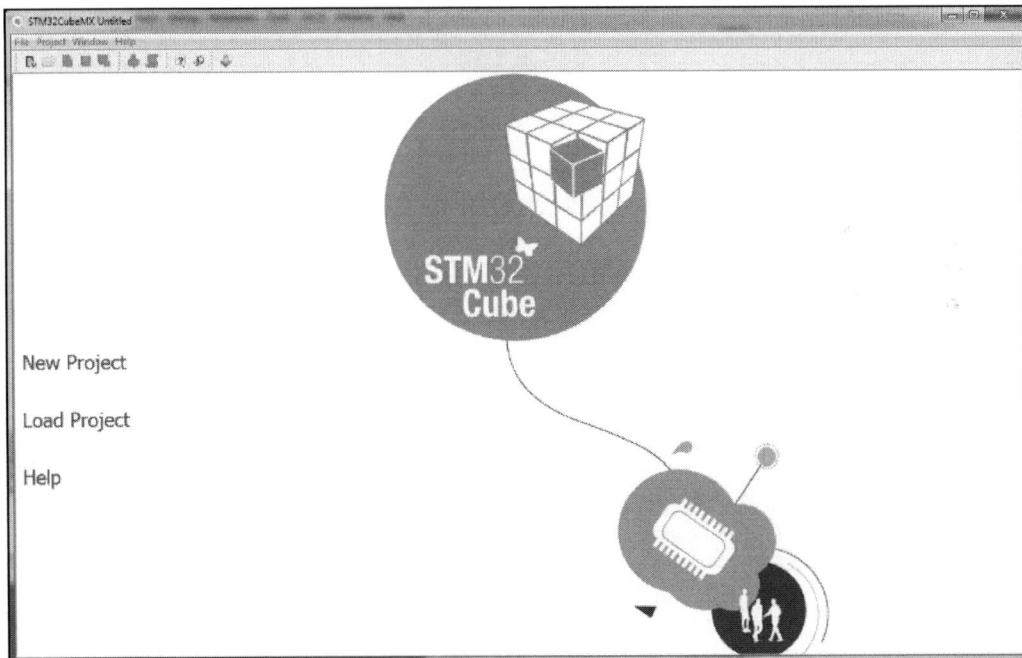

New Project

Load Project

Help

6. You should now see the microcontroller part rendered on the screen, as in the following screenshot:

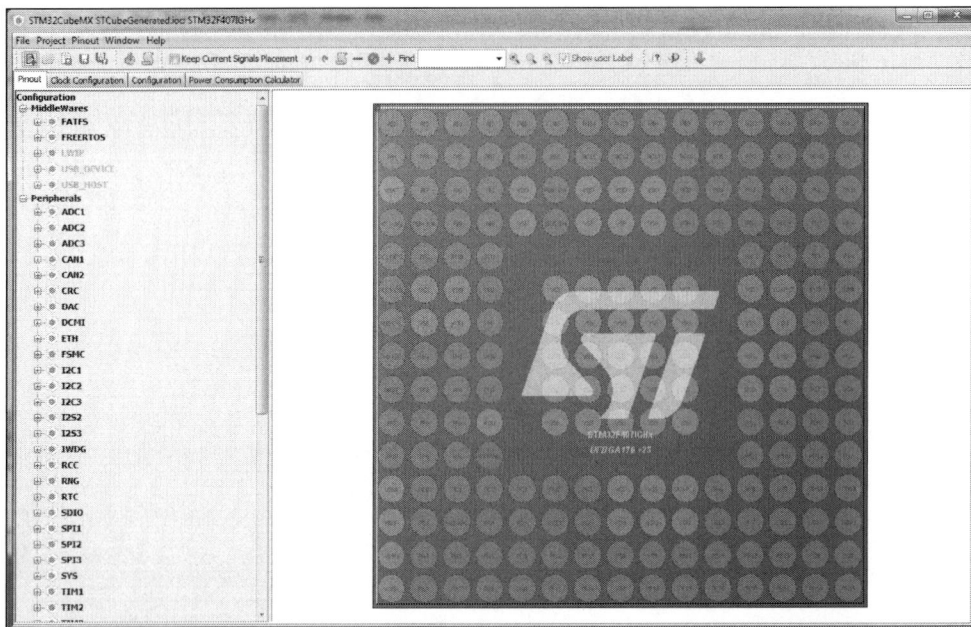

7. Select pin **[G1]** (left mouse button) and use the drop-down menu to configure the pin as **RCC OSC IN**, as in the following screenshot:

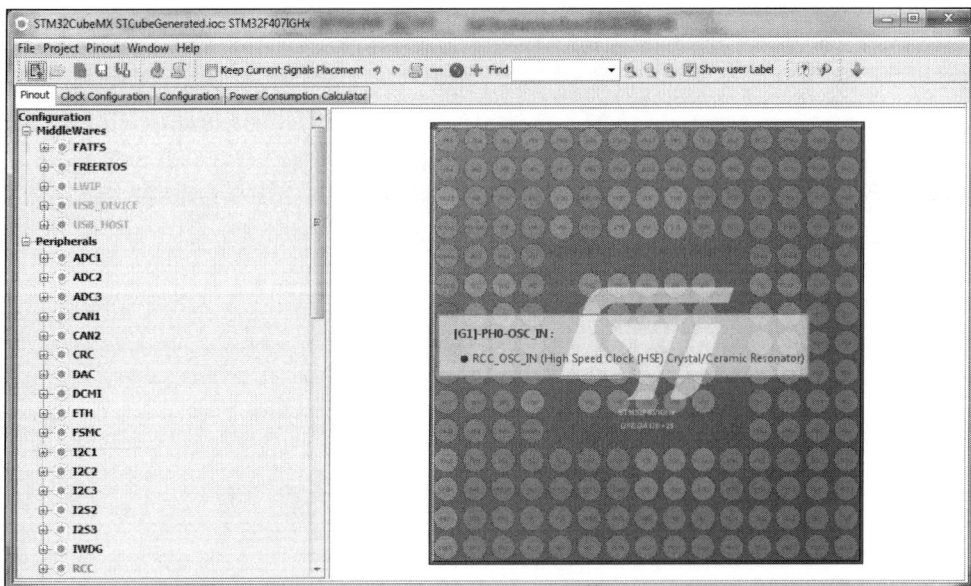

8. Similarly, configure pin **[H1]** as **RCC OSC OUT**.

9. Expand **Peripherals** → **RCC** and use the drop-down menu to configure the **HSE** to use a **Crystal/Ceramic Resonator**:

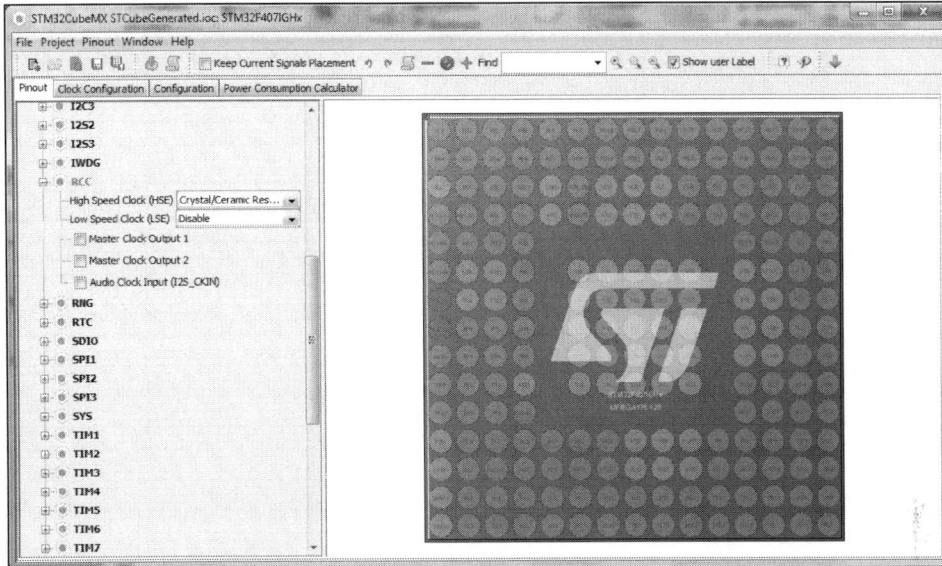

10. Open the **Clock Configuration** tab and configure the clock tree to use a **25** MHz input (crystal), set the clock divider, and select **PLLCLK** to give a **SYSCLK** frequency of **168** MHz. Also, set the **AHB**, **APB1**, and **APB2** Prescalers:

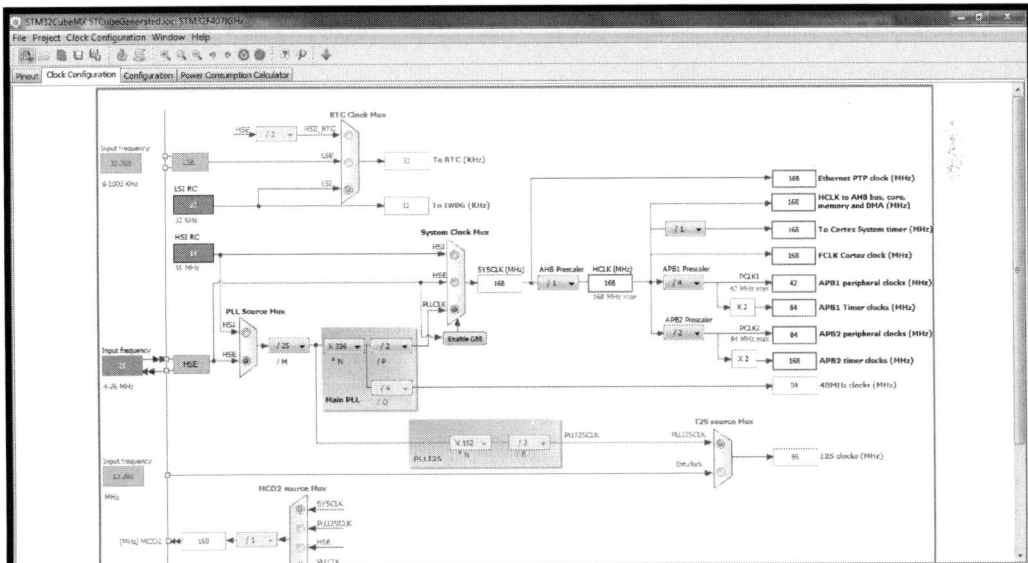

11. Select **Project → Generate Code**.

12. Select **File → Save Project**. Note that the **Toolchain / IDE** is **EWARM**:

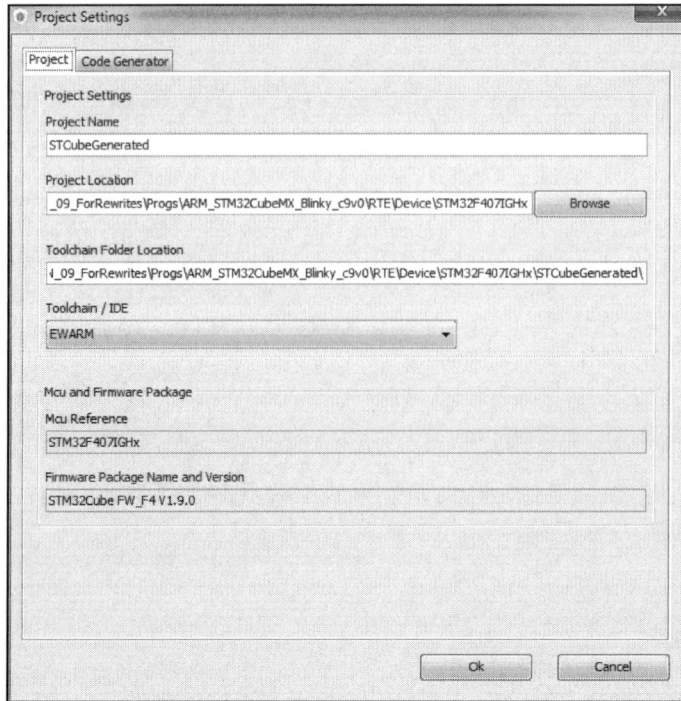

13. Select **OK**; then, quit **STM32CubeMX** by navigating to **File → Exit**.

14. We should see the following message when we return to uVision. Select **Yes** to import the code that we've just generated:

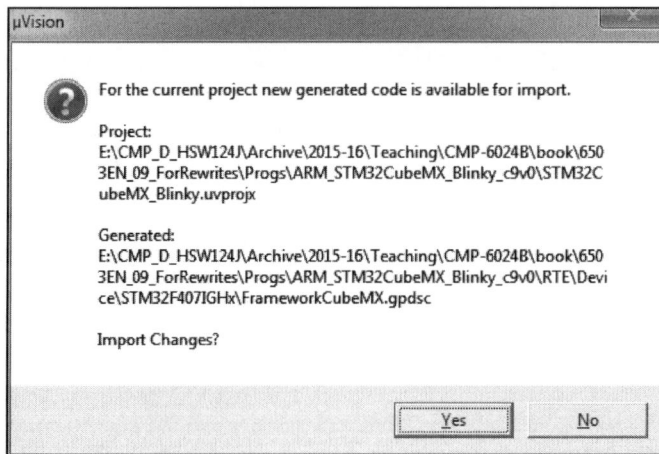

15. Open the **Project** tab and check whether we have successfully imported the code:

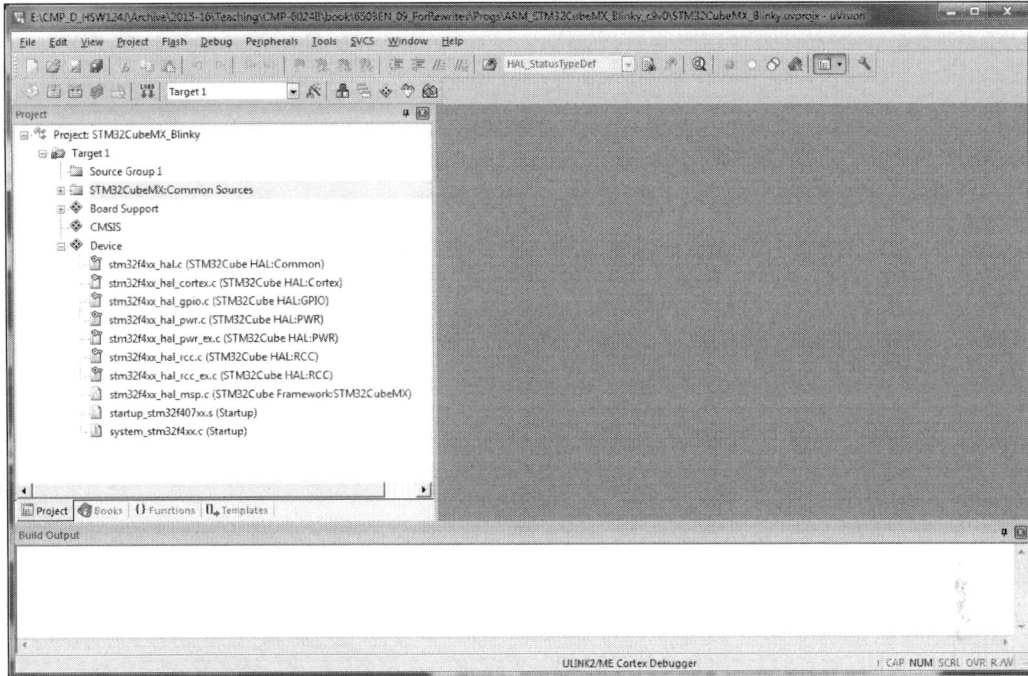

16. Open the file, `main.c` (found in folder `STM32CubeMX:Common Sources`), navigate to the `main()` function definition, and add this statement in the section identified by the `/* USER CODE BEGIN 2 */` comment:

```
LED_Initialize ( );
```

17. Add this code fragment in the section identified by the `/* Infinite loop */` comment:

```
LED_On(0);
for (i=0; i<1000000; i++)
    ;
    LED_Off(0);
for (i=0; i<1000000; i++)
    ;
```

18. Remember to declare the loop variables: i and `#include "Board_LED.h"`.

19. Build, download, and run the program.

How it works...

We've used STM32CubeMX to generate a very basic runtime environment. We're still using the Board Support API to provide functions to configure GPIO and drive LEDs. STM32CubeMX is much more powerful, and we've only illustrated a very basic configuration. More details and further tutorials can be found at `www.st.com`.

There's more...

We can also use STM32CubeMX to configure the GPIO pins that are used to drive the LEDs. We illustrate this in the `ARM_STM32CubeMX_Blinky_c9v1`:

1. After configuring the oscillator (Step 7), select each of the GPIO pins that are connected to the LEDs (**GPIO PG6,7,8, PH2,3,6,7, PI10**) and configure them as outputs, as in the following screenshot:

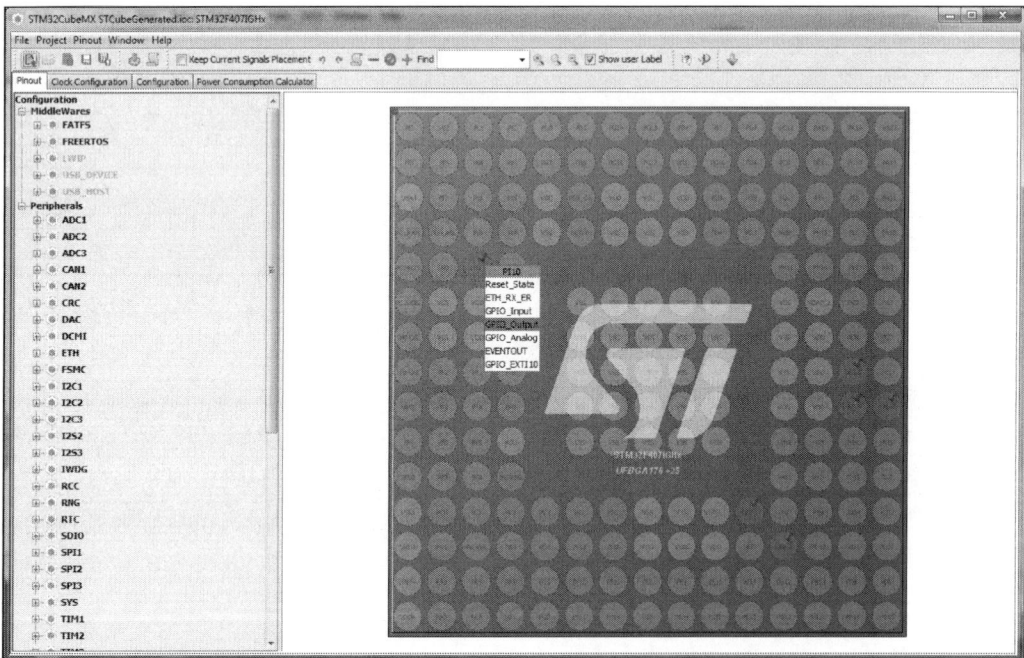

2. Then, select the **GPIO** menu in the configuration tab to set the other GPIO pin parameters (**GPIO Mode**, **Pull-up**, and so on.):

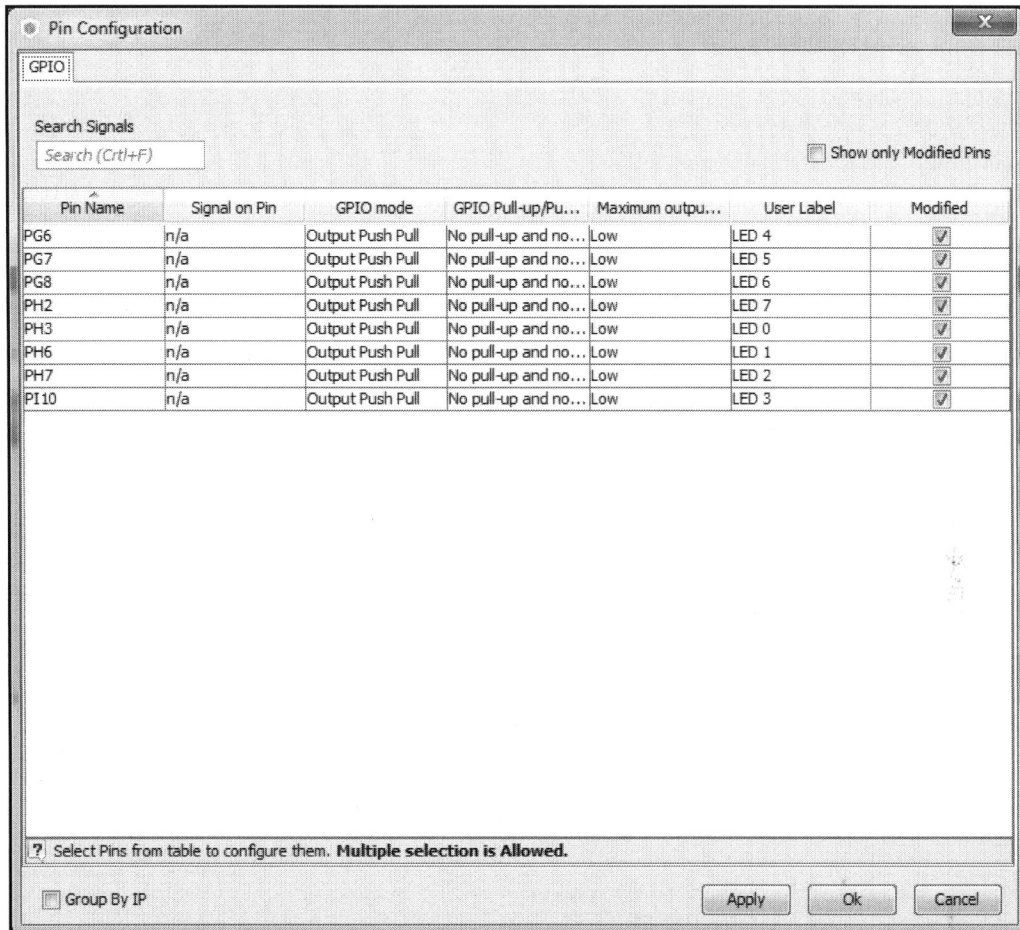

3. Use STMCubeMX, as we did before, to generate the code. When we open the main.c file, we should now find that STM32CubeMX has added code to configure the **GPIO** pins in the MX_GPIO_Init() function, as follows:

```
void MX_GPIO_Init(void)
{

    GPIO_InitTypeDef GPIO_InitStruct;

    /* GPIO Ports Clock Enable */
    __GPIOI_CLK_ENABLE();
```

```
__GPIOH_CLK_ENABLE();
__GPIOG_CLK_ENABLE();

/*Configure GPIO pin : LED_3_Pin */
GPIO_InitStruct.Pin = LED_3_Pin;
GPIO_InitStruct.Mode = GPIO_MODE_OUTPUT_PP;
GPIO_InitStruct.Pull = GPIO_NOPULL;
GPIO_InitStruct.Speed = GPIO_SPEED_LOW;
HAL_GPIO_Init(LED_3_GPIO_Port, &GPIO_InitStruct);

/*Configure GPIO pins : LED_7_Pin
                        LED_0_Pin LED_1_Pin LED_2_Pin */
GPIO_InitStruct.Pin =
        LED_7_Pin|LED_0_Pin|LED_1_Pin|LED_2_Pin;
GPIO_InitStruct.Mode = GPIO_MODE_OUTPUT_PP;
GPIO_InitStruct.Pull = GPIO_NOPULL;
GPIO_InitStruct.Speed = GPIO_SPEED_LOW;
HAL_GPIO_Init(GPIOH, &GPIO_InitStruct);

/*Configure GPIO pins : LED_6_Pin LED_5_Pin LED_4_Pin */
GPIO_InitStruct.Pin = LED_6_Pin|LED_5_Pin|LED_4_Pin;
GPIO_InitStruct.Mode = GPIO_MODE_OUTPUT_PP;
GPIO_InitStruct.Pull = GPIO_NOPULL;
GPIO_InitStruct.Speed = GPIO_SPEED_LOW;
HAL_GPIO_Init(GPIOG, &GPIO_InitStruct);

}
```

4. The `MX_GPIO_Init()` function that was generated by STM32CubeMX is almost identical to that of `LED_Initialize()`. As such, there is no need to call `LED_Initialize ()` before calling `LED_On()` and `LED_Off()`.

How to port uVision projects to GNU ARM Eclipse

STM32CubeMX can also be integrated within the Eclipse IDE and used to configure the RTE in a similar way because it is used by uVision. However, although STM provides a plug-in to invoke STM32CubeMX (refer to *STSW-STM32095* at `www.stm.com`), the current situation is that the code generated is not automatically copied across to the Eclipse project. Luckily, there is a Python v2.7 script called **CubeMXImporter** that allows this to be done easily (note that the procedure is documented at `http://www.carminenoviello.com/`). As Carmine documents this process so thoroughly, this recipe will just explain how to port one of the recipes that we developed earlier in the book. We've chosen `HelloLCD_c2v0` from the *Writing to the GLCD* recipe in *Chapter 2, C Language Programming*, to illustrate this procedure; we call this recipe: `Eclipse_STM32CubeMX_HelloLCD_c9v0`.

How to do it...

1. Follow the instructions at `http://www.carminenoviello.com/` and create a new Eclipse project using the GNU ARM Plugin (that is, navigate to **File → New → C Project**). We'll assume that this project is called `test5`. Use the Hello World ARM Cortex-M C/C++ project template. Note that STM32F407IG has 1024 Kb Flash and 192 Kb RAM.

2. Install and invoke the STM32CubeMX Eclipse plug-in (refer to UM1718 sections 3.2.2 and 3.4.3 at `www.stm.com`). Note that, alternatively, we can run `STM32CubeMX` as a standalone application.

3. Use `STM32CubeMX` to configure and generate code for the `STM32F407IGHX` exactly as we did in the `ARM_STM32CubeMX_Blinky_c9v0` folder in the *How to use the STM32CubeMX Framework*. Note that it's really important to choose SW4STM32 as **Toolchain/IDE** (rather than **EWARM**) before generating the code. Note that I named my `STM32CubeMX` project `mymcu`.

4. Open a command window and run the following:

```
$python cubemximporter.py <path-to-eclipse-workspace>/test5 <path-to-cubemx-out>/mymcu
```

5. We now need to import the **Board Support** to handle the LCD. We can locate the necessary source and include files by right-clicking them in the `HelloLCD_c2v0` folder in the *Writing to the GLCD* recipe in *Chapter 2, C Language Programming*:

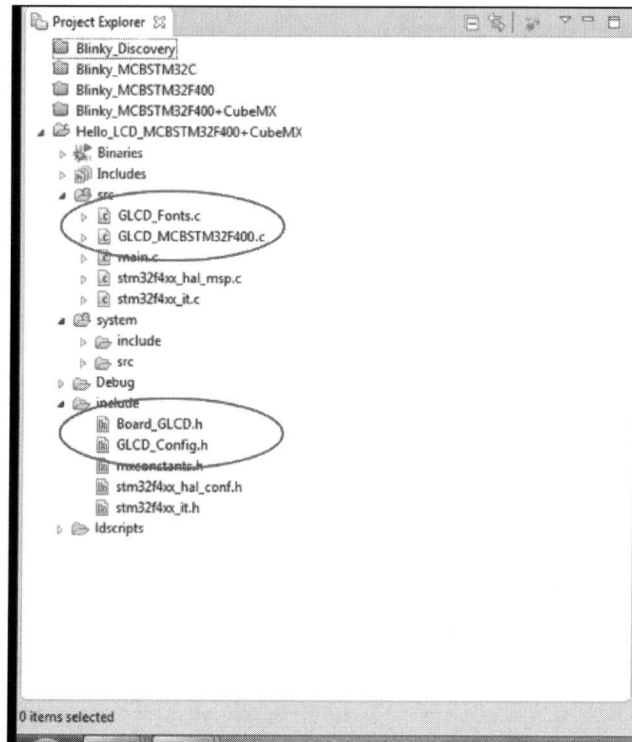

6. Open `main.c` and update `main()` as follows:

```
int main(void)
{

    /* Reset of all peripherals, Initializes the Flash
       interface and the Systick. */
    HAL_Init();

    /* Configure the system clock */
    SystemClock_Config();

    /* Initialize all configured peripherals */
    MX_GPIO_Init();

    /* USER CODE BEGIN 2 */
```

```
GLCD_Initialize();
GLCD_SetBackgroundColor (GLCD_COLOR_WHITE);
GLCD_ClearScreen ();                    /* clear the GLCD */
GLCD_SetBackgroundColor(GLCD_COLOR_BLUE);
GLCD_SetForegroundColor(GLCD_COLOR_WHITE);
GLCD_SetFont (&GLCD_Font_16x24);
GLCD_DrawString(0, 0*24, " CORTEX-M4 COOKBOOK ");
GLCD_DrawString(0, 1*24, "  PACKT Publishing  ");

GLCD_SetBackgroundColor(GLCD_COLOR_WHITE);
GLCD_SetForegroundColor(GLCD_COLOR_BLUE);
GLCD_DrawString(0,3*24,"    Hello LCD    ");
GLCD_DrawString(0,4*24," ARM GNU Eclipse!");

/* USER CODE END 2 */
```

7. Create a **Flash Download Configuration** and flash the program. (Note that **Target Options** invokes uVision5.):

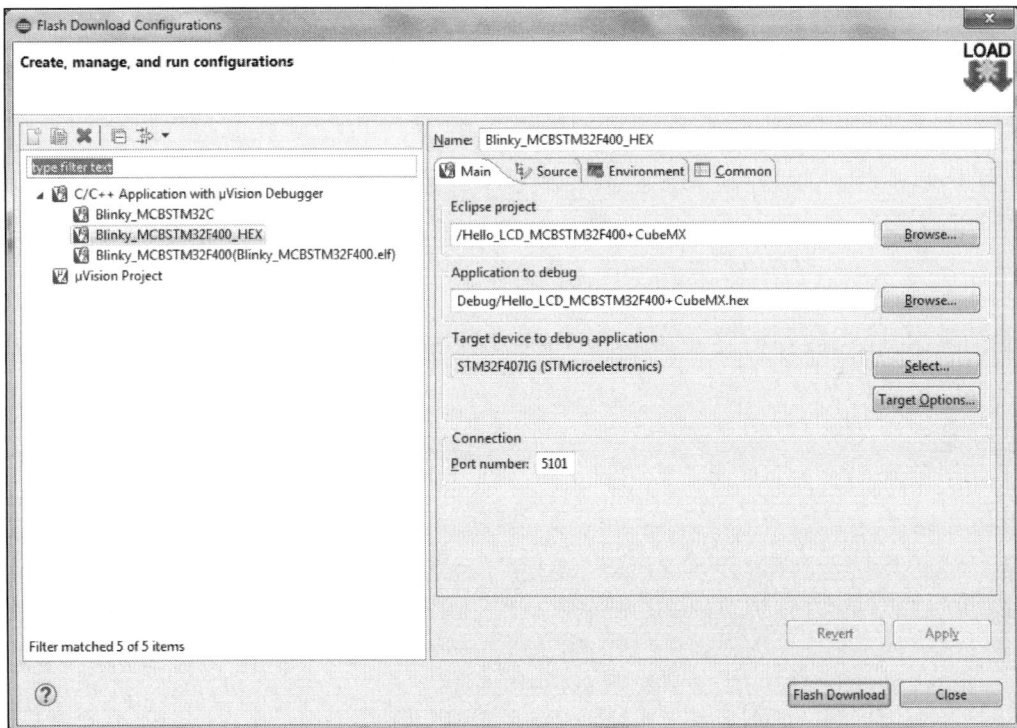

Index

[PACKT] Thank you for buying
PUBLISHING
ARM® Cortex® M4 Cookbook

About Packt Publishing

Packt, pronounced 'packed', published its first book, *Mastering phpMyAdmin for Effective MySQL Management*, in April 2004, and subsequently continued to specialize in publishing highly focused books on specific technologies and solutions.

Our books and publications share the experiences of your fellow IT professionals in adapting and customizing today's systems, applications, and frameworks. Our solution-based books give you the knowledge and power to customize the software and technologies you're using to get the job done. Packt books are more specific and less general than the IT books you have seen in the past. Our unique business model allows us to bring you more focused information, giving you more of what you need to know, and less of what you don't.

Packt is a modern yet unique publishing company that focuses on producing quality, cutting-edge books for communities of developers, administrators, and newbies alike. For more information, please visit our website at www.packtpub.com.

Writing for Packt

We welcome all inquiries from people who are interested in authoring. Book proposals should be sent to author@packtpub.com. If your book idea is still at an early stage and you would like to discuss it first before writing a formal book proposal, then please contact us; one of our commissioning editors will get in touch with you.

We're not just looking for published authors; if you have strong technical skills but no writing experience, our experienced editors can help you develop a writing career, or simply get some additional reward for your expertise.

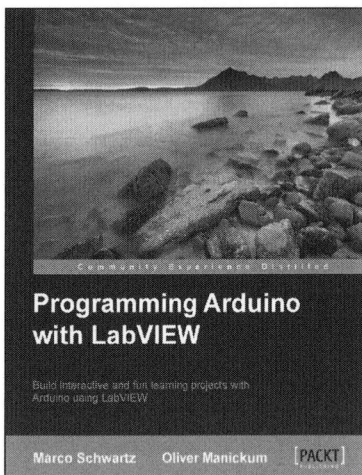

Programming Arduino with LabVIEW

ISBN: 978-1-84969-822-1   Paperback: 102 pages

Build interactive and fun learning projects with Arduino using LabVIEW

1. Learn how to use LabVIEW to control your Arduino projects.

2. Learn how to control a motor from the LabVIEW interface.

3. Automate your Arduino projects without writing a single line of code.

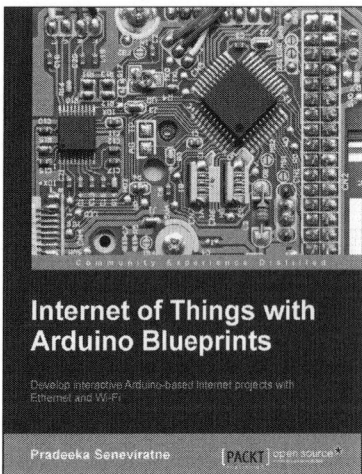

Internet of Things with Arduino Blueprints

ISBN: 978-1-78528-548-6   Paperback: 210 pages

Develop interactive Arduino-based Internet projects with Ethernet and WiFi

1. Build Internet-based Arduino devices to make your home feel more secure.

2. Learn how to connect various sensors and actuators to the Arduino and access data from Internet.

3. A project-based guide filled with schematics and wiring diagrams to help you build projects incrementally.

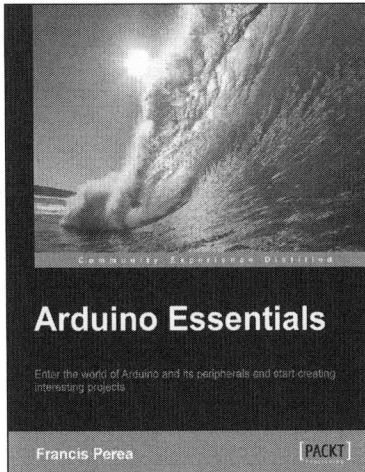

Arduino Essentials

ISBN: 978-1-78439-856-9   Paperback: 206 pages

Enter the world of Arduino and its peripherals and start creating interesting projects

1. Meet Arduino and its main components and understand how they work to use them in your real-world projects.

2. Assemble circuits using the most common electronic devices such as LEDs, switches, optocouplers, motors, and photocells and connect them to Arduino.

3. A Precise step-by-step guide to apply basic Arduino programming.

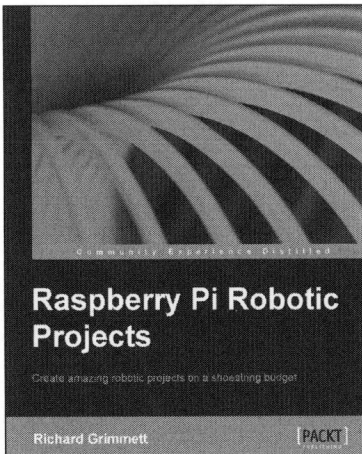

Raspberry Pi Robotic Projects

ISBN: 978-1-84969-432-2   Paperback: 278 pages

Create amazing robotic projects on a shoestring budget

1. Use Raspberry Pi to speak and understand speech.

2. Provide vision capabilities to your robotic projects using a standard webcam.

3. Full of simple, easy-to-understand instructions for bringing your Raspberry Pi on-line for development.

Please check **www.PacktPub.com** for information on our titles

Printed in Great Britain
by Amazon